Final Journeys

Migrant End-of-life Care and Rituals in Europe

Edited by
Alistair Hunter and Eva Soom Ammann

Routledge
Taylor & Francis Group
LONDON AND NEW YORK

First published 2018
by Routledge

2 Park Square, Milton Park, Abingdon, Oxfordshire OX14 4RN
52 vanderbilt Avenue, New York, NY 10017

Routledge is an imprint of the Taylor & Francis Group, an informa business

First issued in paperback 2019

British Library Cataloguing in Publication Data
A catalogue record for this book is available from the British Library

ISBN 13: 978-1-138-74970-2 (hbk)
ISBN 13: 978-0-367-23098-2 (pbk)

Typeset in Minion
by RefineCatch Limited, Bungay, Suffolk

Publisher's Note
The publisher accepts responsibility for any inconsistencies that may have arisen during the conversion of this book from journal articles to book chapters, namely the possible inclusion of journal terminology.

Contents

Citation Information vii
Notes on Contributors ix

1. End-of-life Care and Rituals in Contexts of Post-migration Diversity in Europe: An Introduction 1
Alistair Hunter and Eva Soom Ammann

2. The 'Other' in End-of-life Care: Providers' Understandings of Patients with Migrant Backgrounds 9
Sandra Torres, Pernilla Ågård and Anna Milberg

3. The Art of Enduring Contradictory Goals: Challenges in the Institutional Co-construction of a 'good death' 24
Eva Soom Ammann, Corina Salis Gross and Gabriela Rauber

4. End-of-life Care and Beyond 39
Fuusje de Graaff

5. Between Civil Society and the State: Bureaucratic Competence and Cultural Mediation among Muslim Undertakers in Berlin 53
Osman Balkan

6. The Importance of a Religious Funeral Ceremony Among Turkish Migrants and Their Descendants in Germany: What Role do Socio-demographic Characteristics Play? 68
Nadja Milewski and Danny Otto

7. Staking a Claim to Land, Faith and Family: Burial Location Preferences of Middle Eastern Christian Migrants 85
Alistair Hunter

8. Uncertain Belongings: Absent Mourning, Burial, and Post-mortem Repatriations at the External Border of the EU in Spain 101
Gerhild Perl

Index 117

Citation Information

The chapters in this book were originally published in the *Journal of Intercultural Studies*, volume 37, issue 2 (April 2016). When citing this material, please use the original page numbering for each article, as follows:

Chapter 1

End-of-life Care and Rituals in Contexts of Post-migration Diversity in Europe: An Introduction
Alistair Hunter and Eva Soom Ammann
Journal of Intercultural Studies, volume 37, issue 2 (April 2016), pp. 95–102

Chapter 2

The 'Other' in End-of-life Care: Providers' Understandings of Patients with Migrant Backgrounds
Sandra Torres, Pernilla Ågård and Anna Milberg
Journal of Intercultural Studies, volume 37, issue 2 (April 2016), pp. 103–117

Chapter 3

The Art of Enduring Contradictory Goals: Challenges in the Institutional Co-construction of a 'good death'
Eva Soom Ammann, Corina Salis Gross and Gabriela Rauber
Journal of Intercultural Studies, volume 37, issue 2 (April 2016), pp. 118–132

Chapter 4

End-of-life Care and Beyond
Fuusje de Graaff
Journal of Intercultural Studies, volume 37, issue 2 (April 2016), pp. 133–146

Chapter 5

Between Civil Society and the State: Bureaucratic Competence and Cultural Mediation among Muslim Undertakers in Berlin
Osman Balkan
Journal of Intercultural Studies, volume 37, issue 2 (April 2016), pp. 147–161

Chapter 6

The Importance of a Religious Funeral Ceremony Among Turkish Migrants and Their Descendants in Germany: What Role do Socio-demographic Characteristics Play?
Nadja Milewski and Danny Otto
Journal of Intercultural Studies, volume 37, issue 2 (April 2016), pp. 162–178

Chapter 7

Staking a Claim to Land, Faith and Family: Burial Location Preferences of Middle Eastern Christian Migrants
Alistair Hunter
Journal of Intercultural Studies, volume 37, issue 2 (April 2016), pp. 179–194

Chapter 8

Uncertain Belongings: Absent Mourning, Burial, and Post-mortem Repatriations at the External Border of the EU in Spain
Gerhild Perl
Journal of Intercultural Studies, volume 37, issue 2 (April 2016), pp. 195–209

Notes on Contributors

Pernilla Ågård is a doctoral student at the Department of Sociology, Uppsala University, Sweden.

Osman Balkan is a Visiting Assistant Professor in the Department of Political Science at Swarthmore College, Swarthmore, PA, USA.

Fuusje de Graaff is a Researcher and Trainer at Bureau MUTANT, The Hague, The Netherlands. MUTANT provides training for professionals, managers and users in education and health care.

Alistair Hunter is a British Academy Postdoctoral Fellow at The Alwaleed Centre, University of Edinburgh, UK.

Anna Milberg is an Associate Professor in Geriatrics and Palliative Medicine at the Department of Social and Welfare Studies (ISV), Linköping University, Sweden.

Nadja Milewski is an Assistant Professor of Demography at the Faculty of Economic and Social Sciences, University of Rostock, Germany.

Danny Otto is a doctoral student in the Graduate School "Power of Interpretation. Religion and Belief Systems in Hermeneutic Conflicts" at the University of Rostock, Germany.

Gerhild Perl is a doctoral student at the Institute of Social Anthropology, University of Bern, Switzerland, and a visiting PhD Researcher at the University of Cambridge, UK.

Gabriela Rauber is a doctoral student at the Institute of Social Anthropology, University of Bern, Switzerland.

Corina Salis Gross is an Associate Researcher at the Institute of Social Anthropology, University of Bern, Switzerland.

Eva Soom Ammann is an Associate Researcher at the Institute of Social Anthropology, University of Bern, Switzerland, as well as a Lecturer and Senior Researcher in Nursing at the Health Division of Bern University of Applied Sciences, Switzerland.

Sandra Torres is a Professor in Sociology and Chair in Social Gerontology at the Department of Sociology, Uppsala University, Sweden.

End-of-life Care and Rituals in Contexts of Post-migration Diversity in Europe: An Introduction

Alistair Hunter[a] and Eva Soom Ammann[b]

[a]Islamic and Middle Eastern Studies, Edinburgh, UK; [b]Health Division of the University of Applied Sciences, Bern, Switzerland

This special issue on migrant dying and death showcases the work of a number of authors exploring a newly emerging field of study within European research and policy contexts. Indeed before the 2000s the question of dying and death in migratory contexts received very little attention from researchers, be it in the social sciences or beyond. At first sight this is puzzling, given that when treated separately these fields have constituted a rich terrain for scholarly inquiry, not least in the discipline of sociology where studies of dying and death (Durkheim 1897) and migration (Thomas and Znaniecki 1918, Park 1928) have been foundational. Even social demographers, who by vocation are attentive to mortality as well as population mobility, seemed reluctant to enter this field of study. In the period before 2000, population health was the only scientific domain in which a substantial body of knowledge accumulated, focussing on the apparent paradox of lower than average mortality rates among migrants in industrialised countries (Markides and Coreil 1986, Abraido-Lanza *et al.* 1999). Migration scholarship in the social sciences, by contrast, was oriented to younger people, of working age. In following this orientation, migration studies have arguably internalised the priorities of governmental actors and employers, for whom the costs and benefits of migration have long been measured in economic terms (Sayad 2006). This is particularly the case in Europe since the 1960s, which constitutes the geographical and temporal frame of reference for this special issue. As Berger and Mohr's parody of the prevailing logic of European guestworker capitalism put it: 'So far as the economy of the metropolitan country is concerned, migrant workers are immortal […] they do not age: they do not get tired: they do not die' (Berger and Mohr 1975: 64).

This short-sighted approach began to lose credibility in the late 1980s, as the first generation of post-WWII labour migrants to Europe began to retire from (or were forced out of) the labour market. The first studies on migrant ageing, focusing on countries which recruited migrants early such as Britain, France and Switzerland, were intended to sound the alarm on the poverty and ill-health which touched the ex-migrant workers, many of whom had worked in physically wearing and poorly paid manual jobs (Samaoli 1989, Blakemore and Boneham 1994, Bolzman *et al.* 1996). The first studies on death and dying amongst this pioneer migrant generation would follow shortly, with a number of research monographs (Firth 1997, Tan 1998, Chaïb 2000, Gardner 2002) and vanguard contributions in peer-reviewed journals (see, for example, Jonker 1996, Reimers 1999, Oliver 2004) appearing around 2000.

It is noteworthy that many of these early studies were written by anthropologists or undertaken from an ethnographic perspective, and that most took as their subject Muslim communities in different European countries. Furthermore, they focused primarily on what happens in migrant communities *after* a death occurs, rather than describing the preceding transitions involved in dying. The themes which they treated can be summarised under three headings: identities, rituals and legal–institutional aspects. The question of multiple identities and place attachments looms large: many first-generation migrants strived hard to maintain ties with places of origin, yet the experience of settlement in European countries also strongly marked their worldviews (Reimers 1999, Gardner 2002). Funeral rituals are the last opportunity to express such ties of belonging, and these may be complicated in a migration context. Muslims in particular, because of the religious imperatives of whole body burial and uninterrupted repose, have been confronted with a stark choice about where to be interred. The overwhelming preference reported in these early works is for funeral rituals to take place in countries of origin (Chaïb 2000, Gardner 2002), although opting for countries of immigration is interpreted as a practice which would anchor future generations in Europe (Jonker 1996, Reimers 1999, Chaïb 2000). Related to this, a further theme concerns the legal–institutional barriers which arise when relatives, religious specialists and undertakers in countries of residence attempt to faithfully replicate the end-of-life rituals practised in places of origin (Firth 1997, Chaïb 2000, Gardner 2002). Public authorities have erected a rather strict framework of regulations for the treatment and transport of corpses, which sets clear limits to the freedom of organising burial rituals – especially if they stretch over transnational spaces as is often the case for migrants (Zirh 2012).

Legal–institutional barriers may also arise before the funeral, during the different transitions of dying. This is foregrounded in studies of the palliative care provided to dying migrants (see, Spruyt 1999, de Graaff and Francke 2003, Evans *et al.* 2011, Gunaratnam 2013, Salis Gross *et al.* 2014). Dying in European late modernity is characterised by a shift from sudden death (for example, from accidents, violence or infection) to slower dying caused by chronic disease and which to a certain degree can be managed and planned by specialist institutions, above all by a highly elaborate medical system with impressive powers to maintain and restore life (Walter 2003, Kellehear 2007, Walter 2012). The manageability of dying, however, also demands that numerous decisions be taken, leading to the construction of the ideal–typical autonomous patient who is expected to be in a position to take prospective decisions and, by doing so, to determine his or her dying. Dying in the context of European health services is in this sense dominated by a very specific 'cultural' ideal of self-determined dying. These professional ideals may collide with patients' or relatives' own views of 'good dying': such incompatibilities may arise in many contexts, including but not limited to instances of migrant dying (Gunaratnam 2013).

An additional aspect inherent in the contested notion of 'good death' centres on where death takes place. This has particular significance in migratory contexts: 'good deaths' may be idealised as taking place 'at home' surrounded by loved ones (see for example, Ariès 1981), whereas dying alone or in a foreign or unfamiliar environment may indicate a 'bad death' (Seale 2004). Increasingly however death takes place in institutions of curative or palliative care, where the biomedical aspects of dying can be well-controlled by care professionals, but potentially at a cost to the dignity of the dying person (Kellehear 2007). Producing a 'good death' furthermore involves the appropriate handling of bodies and the social organisation of bereavement rituals after the medically determined moments of death – a dimension often overlooked by public health and health care professionals (Venhorst 2013).

To summarise, a number of contributions about dying and death in migratory contexts have been published in the last 15 years, initially in social anthropology and later in nursing studies. We hesitate however to describe this scientific production as a coherent body of literature because by and large these contributions have not been in dialogue with each other. In part, this can be attributed to language barriers, with an important output in French (see Chaïb 2000, Petit 2002, Aggoun 2006, Lestage 2012) and German (for example, Tan 1998, Salis Gross *et al.* 2014) remaining largely unknown to Anglophone audiences. Nor has there been much dialogue between European work and research conducted in North America or Australasia, where, in addition to a well-established focus on the challenges of palliative care with ethnically diverse patients (e.g. Turner 2002, McNamara 2004), attention has recently turned to the mounting migrant death toll in increasingly securitised border zones (Nevins 2010, Weber and Pickering 2011).

In reviewing the literature on migrant dying, it is clear that – as in life – no death is the same: differences in cause of death, institutional setting (hospital, hospice or at home), policy contexts (international, national and local) as well as the ethnicity, socioeconomic and residency status of the dying person all lead to manifest heterogeneity. There are, in other words, many angles from which to apprehend dying and death in migratory contexts (Gunaratnam 2013). Without pretending to cover all of them, we do aspire with this special issue to make connections between hitherto closed spheres of inquiry in the hope of generating fresh insights. Furthermore we do so in the expectation that questions around migrant dying will only become more pertinent in Europe in the next decades, as the demographic ageing of migrant communities becomes more prevalent (see, Rallu 2016).

Bringing together seven studies reflecting different institutional and (trans)national contexts of migrant dying, our point of departure is that the end of life is a critical juncture in migration and settlement processes, precipitating novel intercultural negotiations which hitherto have not been examined comparatively by scholars. The papers can be broadly categorised under two themes which emerged as central in the above literature review: end-of-life care and end-of-life rituals. A key issue when facing death is the organisation of adequate care for the dying, which may be a challenging task in pluralised settings involving both migrant patients and migrant carers. In the next section we introduce the papers devoted to care issues. We then turn to the papers treating ritual aspects: facing the end of life furthermore involves the practice of rituals in order to make sense of the transition from life to death. As regards both care and ritual contexts, the papers show that the need to reconcile different cultural, religious and administrative norms relating to death is infused with ontological insecurities which may result in new or renewed interrogations of identities and belongings, frequently attended by the need to (re-)negotiate frames of reference.

Dying and end-of-life care

The current state of European research on migrant dying largely treats migrants as a group of patients posing specific challenges to professionals working in institutions of end-of-life care. Migrants' perspectives on dying and end-of-life care, however, have rarely been put centre stage. Research therefore has a tendency to dwell on the specific professional competences needed to deal with migrants' supposedly unique needs.

Critical reflections on professional approaches to migrants' care needs are the starting point for the first contribution of this special issue. Torres, Ågård and Milberg focus on the views and expectations of Swedish end-of-life care providers towards patients with a migrant background. Drawing on data gathered in focus group interviews with end-of-life care professionals, their paper finds that professionals' views on migrant patients are closely linked to expectations of difference and, by extension, expectations of difficulties. Resolving such difficulties requires specific skills (referred to as 'culture competence'), which they do not think they possess. In their analysis, the authors connect the 'othering' concept of 'culture competent care' to the professionally very well-established concept of 'patient centeredness'. They point to the fact that assuming 'otherness' when dealing with migrant patients at the same time implies the assumption of 'sameness' with respect to non-migrant patients. Patient centeredness, in contrary, stresses the uniqueness of every patient, which in itself contradicts the concept of 'sameness'. In recommending that nursing staff focus on individual uniqueness instead of supposed 'otherness', the authors conclude with a conceptual solution to the practical problems perceived by end-of-life care professionals vis-à-vis patients with migrant backgrounds.

While the first paper focuses on carers, the next contribution by Soom Ammann, Rauber and Salis Gross looks at interactions between carers and dying persons with a migrant background. Based on ethnographic research in Swiss nursing homes focussing on how institutions 'do death', they explore how end-of-life care for migrant elders in long-term care establishments is subject to interpretations by staff of how a 'good death' can be achieved. These interpretations, however, may collide with the interpretations of additional actors involved in 'doing death', such as relatives, but also other professional carers. Caring for dying migrants may involve more diversified ideas, and therefore collisions may arise more readily. They are, however, not clearly traceable to 'cultural' systems of reference in the sense of migrant origin, but are more complex. Thus migrant background is, as has already been argued in the first paper, not a decisive line of difference, but an additional frame of reference requiring consideration.

The third paper, by de Graaff, extends the scope of the special issue in two ways. First, it is inclusive of migrant perspectives by foregrounding relatives' views on end-of-life care provision. The paper draws on data from focus groups with the relatives of deceased cancer patients of Turkish and Moroccan background in the Netherlands, as well as interviews with their palliative care professionals. In resonance with the foregoing papers, the article points to challenges stemming from different implicit normative positions. Second, de Graaff is exploring a genuinely new analytic terrain in focusing on the transition between end-of-life care and the 'burial care' involved in preparing the body for the funeral and accompanying the bereaved. The way relatives experience this transition points to a change in their subjectively experienced positionality, which the author traces back to different logics of care. De Graaff argues that palliative care givers, whose worldview is guided by professional standards, can learn from the customer-oriented sensitivity to diverse needs shown by undertakers.

Death and end-of-life rituals

De Graaff's contribution fruitfully bridges the conceptual gap between end-of-life care and end-of-life rituals and sets the scene for the fourth paper, by Balkan. His focus is the involvement of Muslim undertakers in the intercultural negotiations which are precipitated by the death of Muslims in Berlin, Germany. Drawing on data from interviews and ethnographic research set within Berlin's Islamic funeral and repatriation industry, Balkan pays particular attention to the mediating position which Muslim undertakers have assumed between the German state and Muslim migrant families, mainly of Turkish background. Vis-à-vis the German state, he finds that undertakers are sometimes called to act as unofficial representatives of their communities in order to allay bureaucrats' misconceptions and prejudice about Muslims. Vis-à-vis Muslim migrant families, the undertakers' mastery of state regulations around burial is a key source of their legitimacy as members of a profession which is otherwise stigmatised due to popular beliefs about profiting from others' misfortune. This mastery also sets them apart from their clients, whom some of the interviewed undertakers berate for their lack of 'integration' and inability to understand the rational–legal order of German society.

The next article, by Milewski and Otto, complements Balkan's fine-grained ethnographic study by maintaining the geographical focus on Muslims in Germany. However, Milewski and Otto add a new dimension insofar as their paper is one of the few in this field to adopt a quantitative socio-demographic approach (see also Attias-Donfut and Wolff 2005). More particularly, their contribution is oriented to the literature on religiosity among migrants and their descendants. While quantitative analyses have been undertaken to test competing hypotheses about declines in religiosity due to assimilation or maintenance of religiosity as a means to ethnic identity formation, none have inquired about one key aspect of religiosity, namely religious funerals. Drawing on a sample of 4000 people of Turkish migration background, Milewski and Otto apply logistic regression methods to reveal the significance of different factors in attitudes to religious funerals, such as education, citizenship and partner's origin. They find that the importance of a religious funeral remains high across first and second generations, as well as among those who do not place importance on religion in other phases of the life course.

The sixth paper, by Hunter, continues the discussion initiated by Milewski and Otto regarding the importance of funeral rituals, specifically burial practices. His study, based on qualitative interviews conducted in Britain, Denmark and Sweden, breaks new empirical ground by soliciting the views of Christians of Middle Eastern origin, a migrant group rarely considered by migration scholars working in Europe. Hunter's paper examines how identities are negotiated in and through place by posing the question of preferred burial location. In an ever more mobile world, the ability to develop meaningful identifications with multiple places continues up until the end of life, and for migrants this may present a stark choice: to be buried in the place of origin, the place of residence, a third country or the transnational solution of performing rituals in more than one location. In presenting a typology of motivations for preferred place of burial, supported by examples from his interview data, Hunter shows the multiple, nuanced and sometimes ambivalent relationships between place and identity which are negotiated by Middle Eastern Christians living in Europe.

This theme of multiple and ambivalent belongings in the context of death is taken up in the final paper by Perl. Her analysis reveals an important but easily overlooked point, namely that establishing the identity and belonging of the deceased is a crucial pre-condition of appropriate ritual practice at the end of life. Establishing belonging becomes particularly difficult in the context covered by Perl, namely the deaths at sea of illegalised migrants making the journey from the southern shores of the Mediterranean to Europe. Her article thus extends the empirical scope of the special issue by shining light on a phenomenon which, distressingly and shamefully, has been an all too common occurrence in recent times. Perl goes on to explore the nature of the interactions and power dynamics between the different actors implicated in the identification process – from gravediggers to bureaucrats, police officers, religious institutions and associations of the bereaved – demonstrating how the uncertain belongings of the deceased are produced, suppressed or rewritten. Echoing Balkan's findings about cultural mediation as a means of deflecting the stigma attached to the profession of undertaking, Perl closes her paper by reflecting on the efforts of one Spanish undertaker to identify and repatriate dead border-crossers to Morocco.

By drawing together in this special issue different studies dealing with dying and death in contexts of post-migration diversity in Europe, we hope to initiate a fruitful and more joined-up discussion to take forward this demanding but also very rich field of study. Studying dying and death is in our view not only enriching for migration studies, but also sheds light on dying and death as issues of relevance for humanity as such. The research reported in this issue underlines that migrancy – both as an administrative status and as a biographical set of experiences – has the potential to unsettle established norms surrounding dying and death. In this sense, studying dying and death in migratory contexts not only illuminates our understanding of 'the other' but also of 'the self', while at the same time illustrating how diverse lived experiences are as much *within* 'cultures' as *across* them. Particularly enriching is the fact that this special issue is not focusing on the moment of death, but is attempting to bridge the issue by looking back at conditions of dying under the dominant 'cultural' ideal of biomedical palliative care and projecting forward to regulations and practices associated with the handling of corpses and the social organisation of bereavement. Intercultural negotiations around dying and death involving migrants in Europe thus shed new light on the structural inequalities which are inherent in the relationship between migrants and powerful institutions of governance.

Acknowledgments

We are particularly grateful to Marina Richter for her comments on an earlier version of this text.

Disclosure statement

No potential conflict of interest was reported by the authors.

References

Abraido-Lanza, A.F., *et al.*, 1999. The Latino mortality paradox: a test of the "salmon bias" and healthy migrant hypotheses. *American journal of public health*, 89 (10), 1543–1548.

Aggoun, A., 2006. *Les musulmans face à la mort en France*. Paris: Vuibert.

Ariès, P., 1981. *The hour of our death*. London: Allen Lane.

Attias-Donfut, C. and Wolff, F-C., 2005. The preferred burial location of persons born outside France. *Population English edition*, 60 (5–6), 699–720.

Berger, J. and Mohr, J., 1975. *A seventh man: a book of images and words about the experience of migrant workers in Europe*. Harmondsworth: Penguin.

Blakemore, K. and Boneham, M., 1994. *Age, race and ethnicity: a comparative approach*. Buckingham: Open University Press.

Bolzman, C., Fibbi, R., and Vial, M., 1996. La population âgée immigrée face à la retraite: problème social et problématiques de recherche. *In*: H.-R. Wicker, J.-L. Alber, C. Bolzman, R. Fibbi, K. Imhof and A. Wimmer, eds. *Das Fremde in der Gesellschaft : Migration, Ethnizität und Staat = L'altérité dans la société: migration, ethnicité, état*. Zürich: Seismo, 123–142.

Chaïb, Y., 2000. *L'émigré et la mort: la mort musulmane en France*. Aix-en-Provence: Edisud.

Durkheim, É., 1897. *Le Suicide : étude de sociologie*. Paris: F. Alcan.

Evans, N., *et al.*, 2011. Appraisal of literature reviews on end-of-life care for minority ethnic groups in the UK and a critical comparison with policy recommendations from the UK end-of-life care strategy. *BMC health services research*, 11, 141. Available from: http://bmchealthservres.biomedcentral.com/articles/10.1186/1472-6963-11-141.

Firth, S., 1997. *Dying, death and bereavement in a British Hindu community*. Leuven: Peeters Publishers.

Gardner, K., 2002. *Age, narrative and migration: the life course and life histories of Bengali elders in London*. Oxford: Berg.

de Graaff, F.M. and Francke, A.L., 2003. Home care for terminally ill Turks and Moroccans and their families in the Netherlands: carers' experiences and factors influencing ease of access and use of services. *International journal of nursing studies*, 40, 797–805.

Gunaratnam, Y., 2013. *Death and the migrant: bodies, borders and care*. London: Bloomsbury.

Jonker, G., 1996. The Knife's Edge: Muslim burial in the diaspora. *Mortality*, 1 (1), 27–43. doi:10.1080/713685827

Kellehear, A., 2007. *A social history of dying*. Cambridge: Cambridge University Press.

Lestage, F., 2012. Éditorial: La mort en migration. *Revue européenne des migrations internationales*, 28 (3), 7–12.

Markides, K.S., and Coreil, J., 1986. The health of Hispanics in the southwestern United States: an epidemiologic paradox. *Public health reports*, 101 (3), 253–265.

McNamara, B., 2004. Good enough death: autonomy and choice in Australian palliative care. *Social science and medicine*, 58, 929–938.

Nevins, J., 2010. *Operation gatekeeper and beyond: the war on 'illegals' and the remaking of the U.S.-Mexico boundary*. 2nd ed. New York: Routledge.

Oliver, C., 2004. Cultural influence in migrants' negotiation of death. The case of retired migrants in Spain. *Mortality*, 9 (3), 235–254.

Park, R.E., 1928. Human migration and the marginal man. *American journal of sociology*, 33 (6), 881–893.
Petit, A., 2002. L'ultime retour des gens du fleuve Sénégal : Retours d'en France. *Hommes et migrations*, 1236, 44–52.
Rallu, J.-L., 2016. Projections of older immigrants in France, 2008–2028. *Population, space and place*. doi:10.1002/psp.2012
Reimers, E., 1999. Death and identity: graves and funerals as cultural communication. *Mortality*, 4 (2), 147–166. doi:10.1080/713685976
Seale, C., 2004. Media constructions of dying alone: a form of "bad death". *Social science & medicine*, 58 (5), 967–974.
Salis Gross, C., *et al.*, 2014. *Chancengleiche palliative care. Bedarf und Bedürfnisse der Migrationsbevölkerung in der Schweiz*. Muenchen: AVM.
Samaoli, O., 1989. Un autre regard sur les Maghrébins âgés. Les immigrés vieillissent aussi. *Revue hommes et migrations*, 1126, 15–24.
Sayad, A., 2006. *L'immigration ou Les paradoxes de l'alterité : Tome 1, L'illusion du provisoire*. Paris: Raisons d'agir.
Spruyt, O., 1999. Community-based palliative care for Bangladeshi patients in east London. Accounts of bereaved carers. *Palliative medicine*, 13, 119–129.
Tan, D., 1998. *Das fremde Sterben: Sterben, Tod und Trauer unter Migrationsbedingungen*. Frankfurt am Main: IKO-Verlag.
Thomas, W.I. and Znaniecki, F., 1918. *The Polish peasant in Europe and America: monograph of an immigrant group*. Vol. 1. Primary-group organization. Chicago: University of Chicago Press.
Turner, L., 2002. Bioethics and end-of-life care in multi-ethnic settings: cultural diversity in Canada and the USA. *Mortality*, 7, 285–301.
Venhorst, C., 2013. *Muslims ritualising death in the Netherlands: death rites in a small town context*. Münster: LIT Verlag.
Walter, T., 2003. Historical and cultural variants on the good death. *BMJ British medical journal*, 327, 218–220.
Walter, T., 2012. Why different countries manage death differently: a comparative analysis of modern urban societies. *The British journal of sociology*, 63, 123–145.
Weber, L. and Pickering, S., 2011. *Globalization and borders: death at the global frontier*. Basingstoke: Palgrave Macmillan.
Zirh, B., 2012. Following the dead beyond the 'nation': a map for transnational Alevi funerary routes from Europe to Turkey. *Ethnic and racial studies*, 35, 1758–1774.

The 'Other' in End-of-life Care: Providers' Understandings of Patients with Migrant Backgrounds

Sandra Torres[a], Pernilla Ågård[a] and Anna Milberg[b]

[a]Department of Sociology, Uppsala University, Uppsala, Sweden; [b]Department of Advanced Home Care and Department of Social and Welfare Studies, Linköping University, Linköping, Sweden

ABSTRACT

Research on how end-of-life care providers make sense of cultural, ethnic and religious diversity is relatively scarce. This article explores end-of-life care providers' understandings of patients with migrant backgrounds through a study based on focus group interviews. The analysis brings to the fore three themes: the expectation that the existence of difference and uncertainty is a given when caring for patients with migrant backgrounds; the expectation that the extension of responsibility that difference entails creates a variety of dilemmas; and the expectation that difference will bring about misunderstandings and that patients' needs can go unmet as a result of this. On the basis of these themes we suggest that the end-of-life care providers interviewed regard patients with migrant backgrounds as 'Others' and themselves as providers that cannot deliver so called culture-competent care. The findings are problematised using the lens that the debate on patient-centredness offers. The article suggests that if the uniqueness of all patients is to be seriously taken into account then 'Othering' is perhaps what patient-centredness actually entails.

An increasing number of people around the world have a migrant background (United Nations 2013). The number of people who spend their old age far away from the countries they once called home has also increased over the past few decades (Warnes *et al.* 2004). This means that an increasing number of end-of-life care providers – which is the term we hereby use to refer to health-care professionals who care for dying patients – must nowadays engage in cross-cultural interaction (see Johnstone and Kanitsaki 2009, Evans *et al.* 2012, Gysels *et al.* 2012). By cross-cultural interaction we mean interaction that takes place across cultural, ethnic, religious and language boundaries. It is against this backdrop that the debate on culture-competent end-of-life care started a decade ago when researchers in this field began to advocate for increased culture-competence among care providers (see Doorenbos and Schim 2004). Not all end-of-life care scholars believe, however, that the adoption of culture-competent models in this specific care setting is what is needed.

Some problematise the simplified ways in which these models approach the question of cross-cultural care interaction because they can lead to the culturalisation of patients (see Molassiotis 2004) and can end up reinforcing the very barriers that they are meant to remove. Here the term 'culturalisation' is used to allude to the essentialisation that takes place when stereotypical assumptions about patients' ethno-cultural backgrounds are used as a compass to guide the care they receive.

Jones (2005) argues for example that end-of-life care providers should be sceptical of the cookbook-like approaches that are characteristic of these models. She urges end-of-life care providers to be aware of the fact that the simple guidelines for so-called culture-competent approaches to various ethnic groups' special needs and end-of-life preferences may create a multitude of myths about patients with ethno-cultural minority backgrounds. Epner and Baile (2012) have argued, in turn, that the literature on culture-competence focuses on 'the categorical', which is why this approach can lead to 'stereotypical thinking rather than clinical competence' (Epner and Baile 2012: iii34). Gunaratnam (2007) has problematised these models because they tend to downplay the challenges that cross-cultural interaction actually entails. In this respect, Perloff *et al.* (2006) argue that the greatest problem with this type of interaction is that it involves meetings between groups that lack confidence in each other, something that these models seldom take into consideration.

Empirical research on cross-cultural interaction within the context of end-of-life care is still scarce. The little research that has looked into the way in which health-care providers experience cross-cultural interaction comes from other care settings. Kai *et al.* (2007) have studied, for example, how this interaction can be experienced in primary and secondary care settings in England, and found that care staff regard this type of interaction as challenging and feel they lack the competence to handle it. Their study shows that there seems to be some apprehension among health-care providers regarding cross-cultural interaction. It does not, however, focus on end-of-life care – a care context often guided by the palliative care philosophy which implicitly suggests that culture-competence and knowledge of religions is necessary if cross-cultural interaction is to function in an optimal manner (see Gysels *et al.* 2013). To this end it must be noted that discussions about what is considered to be optimal in terms of the interaction that takes place within care settings between patients and care providers are nowadays conducted on the basis of the notion of patient-centredness. Research on what end-of-life care providers expect of cross-cultural interaction is, however, needed if we are to understand what is deemed to be challenging in terms of patient-centredness within this specific care context. This article addresses this very angle by exploring the understandings that end-of-life care providers articulate when they talk about the challenges associated with cross-cultural interaction involving patients with migrant backgrounds.

Patient-Centredness and End-of-Life Care

Balint (1955) is credited with having introduced the concept of 'patient-centred medicine' into the medical literature in order to offer an alternative to the 'illness-centredness' that characterised the field of medicine at that point in time. This notion has its roots within the paradigm of holism, which suggests that there is more to a patient than meets the eye and that an understanding of patients' bio-psychosocial entireties is needed if health-care providers are to be able to meet patients' needs in a satisfactory manner.

In the past few decades, patient-centred care has become the goal for many health-care systems around the world. However, there is little consensus as to what patient-centredness actually means, which is why a variety of systematic reviews have been conducted in the past few years (see Kitson *et al.* 2013). Scholl *et al.* (2014) have found 15 different dimensions of patient-centredness which ranged from the idea that the patient is a unique person, to the essential characteristics of clinicians. There are, in other words, numerous ways in which the notion of patient-centredness is used within health-care research to draw attention to the fact that the treatment of illness and disease should not be delivered in a generic manner. Instead, health-care providers should aim to place patients' unique characteristics and needs at the forefront of their treatment. Patient-centredness is therefore about keeping in mind the individuality of each patient; a fact that is rather obvious but that seems to be disregarded more often than we care to admit (Scholl *et al.* 2014).

The notion of patient-centred end-of-life care has received relatively little attention in the health professional literature. Following the growing momentum of the debate on culture-competent care, however, this situation is beginning to change. The principles of palliative care which are endorsed by the WHO emphasise patient/family centredness (in addition to quality of life and the team approach) (Dunn and Miller 2014). According to Gysels *et al.* (2013), most health-care professionals involved in end-of-life care regard the WHO definition of palliative care as their 'master definition' of what end-of-life care should entail, even though they do not agree on what characterises end-of-life care in practice. Thus, although not always as explicitly as one may expect, there is an expectation that end-of-life care should be delivered in a patient-centred way (see Lavoie *et al.* 2013). With respect to the group of patients with which this paper is concerned – patients with migrant backgrounds – Selman *et al.* (2014) have argued that:

> Holistic models of patient care are essential to the practice of patient-centred care. However, these models have up to now largely neglected the role of culture and the search for meaning in the illness experience, despite evidence of disparities in the access of palliative care services by people from ethnic minority groups. (80)

Epner and Baile (2012) have also drawn attention to the specific challenges that health-care providers face when delivering care to patients who have ethno-cultural backgrounds different from their own – when cross-cultural interaction is at stake. They have argued that culture-competent care can only be provided if we depart from patient-centredness. However, these two studies are not the norm since the debate on patient-centredness has yet to be informed by the debate on culture-competent care. The data that have been collected in the project we will report on shortly suggests that end-of-life care providers' understandings of patients with migrant backgrounds offer insights into what they deem to be needed (or missing) if they are to address these patients' needs in a patient-centred way.

Methods

This article is based on data from 11 focus group interviews with health-care professionals ($n = 60$) who provide care for dying patients at 11 different health-care units in Sweden (insight into this specific national context is provided in the next section). The number of participants in the focus groups ranged from three to seven. The data come from a project that aims to explore the ways in which cross-cultural interaction is understood by end-of-life care providers. By 'understandings' we mean the ways in which care professionals regard cross-cultural end-of-life care interaction, irrespective of whether or not they have actually experienced it. Thus, the ontological and epistemological basis for this project is the paradigm that Lincoln and Guba first called 'naturalistic inquiry' but which they have relabelled as constructivism (Guba and Lincoln 1998).

In order to get a wide range of understandings about how cross-cultural interaction is regarded by care professionals involved in end-of-life care, we used a combination of maximum-variation sampling and convenience sampling (see Patton 2002). Those who agreed to be interviewed were asked to fill out a form requesting general information about them before the interview started in order to get insight into their backgrounds. From these forms we learned that 58 out of 60 of our informants were female and worked as nurses or nurse assistants. The health-care units in which they worked had a variety of focuses (internal medicine ($n = 2$), surgery ($n = 3$), geriatrics ($n = 2$), specialised palliative home care ($n = 2$), primary care ($n = 2$)), but all had to provide care to dying patients as one of their assignments. The median number of years of experience in the profession was 16, and within end-of-life care it was 17. There was no requirement that the participants should have their own experience of cross-cultural interaction since the project focuses on care professionals' understandings of this phenomenon as opposed to their experiences of it. The vast majority of our informants (45 out of 60) had, however, some experience of cross-cultural interaction since they had cared for patients with migrant backgrounds (only four said that they had extensive experience of this; 10 said that they had none and 1 failed to respond). One of the Regional Boards of Ethics in Sweden approved the project (Dnr 2011/341-31).

The focus group interviews were semi-structured but used an interview guide (Morgan 1997). Most of the questions posed were open-ended and probed for a variety of issues related to the provision of cross-cultural end-of-life care (e.g. what comes to mind when you think of caring for dying patients with migrant/ethno-cultural minority backgrounds?). Since there is no agreed definition of end-of-life care in practice (see Gysels *et al.* 2013), we used two approaches to arrive at the focus group participants' understandings of this care. First, we posed questions concerned with time, focusing on cross-cultural interaction using a chronological design, for example: 'what comes to mind when you think of the first meeting with the patient?'; '... the further care of the patient?'; '... the care during the very last phase of the patient's life?' '... when the patient dies?'; '... the care after the patient has died'). Second, we approached the end-of-life care angle of their work by using WHO's definition of palliative care (e.g. 'We would now like you to share your thoughts with us about caring for dying patients with migrant backgrounds in relation to WHO's definition of palliative care, where symptom relief in a broad sense is included, as well as communication, teamwork and family'). Follow-up questions were asked when needed. The focus group interviews were transcribed verbatim in order to allow for systematic analysis. During the interviews, dialogical validation of the informants' statements was performed by rephrasing and checking that the statements had been correctly understood. The development of the first preliminary coding scheme which guided the analysis was mainly carried out by the second author using the computer software NVivo 10. This coding scheme – and the content analysis it ended up generating – was examined by the other two authors through peer-debriefing sessions (Creswell 1998), the aim of which was to ensure the trustworthiness of the analysis. The analysis generated a variety of themes which shed light on the challenges that the interviewed care providers associated with cross-cultural interaction in end-of-life care. In this article we explore the care providers' articulations about patients with migrant backgrounds when discussing these challenges.

Sweden as the Context for the Study

Cross-cultural interaction within health-care settings is on the agenda of Swedish care planners and providers due to the ethno-cultural diversity of the Swedish population. During 2012 (the year when the data was collected for this project), Sweden had 9,555,893 inhabitants, of which 15.4 per cent (1,473,256) had a foreign-born background (Statistics in Sweden 2013). More than half of the foreign-born come from a European country. The most common country of birth amongst those with a foreign-born background is Finland (about 11 per cent), and thereafter come Iran and Poland (Statistics in Sweden 2013). It is this diversity and the fact that research on health disparities in this country has shown that people with foreign-born backgrounds have poorer self-reported health than native-born individuals (Wiking, Johansson and Sundqvist 2004) that has put the topic of cross-cultural interaction on the agenda of this country's health-care sector. The National Board of Health and Welfare (2006) stresses, for example, that the ethno-cultural diversity of Sweden's population needs to be taken into account when end-of-life care is being planned and delivered. This report even stresses – albeit in passing – that the diversity in question is best addressed if end-of-life care is provided in a culture-competent manner. It is against this backdrop that we deemed the context in question to be a suitable site for a study that aims to explore end-of-life care providers' understandings of cross-cultural interaction.

We use the term 'patients with migrant backgrounds' in this article because this is the best translation we have for the Swedish word (*invandrarbakgrund*) that is most often used when speaking of people with a foreign-born background in Sweden. Thus, although we are aware that the term 'migrant' is often used in other parts of the world to refer to people who have recently migrated, this is not the way in which the term is used in this country since one can be regarded as a person with a migrant background in Sweden irrespective of how long one has lived in this country or whether or not one has Swedish citizenship.

Anticipating Cross-Cultural Interaction to be Challenging

As already stated, the focus group interviews tended to focus on the challenges that the end-of-life care providers associated with cross-cultural themes. Although most of the end-of-life care providers interviewed (45 out of 60, to be exact) had relatively little experience of this type of interaction, they had numerous ideas about what made these interactions 'different' from what they were used to. In addition, they had rather clear ideas as to what was characteristic of patients with migrant backgrounds, even though few had much experience of either cross-cultural interaction in this setting or the provision of end-of-life care to patients with migrant backgrounds (see methods section for sample characteristics). In almost all of the focus groups, statements were made such as: 'In all fairness, I don't actually know since I don't have that much experience but I expect that it will be ... '; 'I don't have first-hand experience of dealing with these patients but I have heard that ... ', or 'We haven't cared for that many immigrants in our ward but we take for granted that ... ', but the discussions were carried on as if they all knew what they were talking about. This is something that needs to be kept in mind since what this article taps into is what seems to be at the core of the focus group discussions about cross-cultural interaction involving patients with migrant backgrounds, namely the assumption that 'difference' poses a challenge to the delivery of patient-centred care.

Expecting Difference = Feeling Uncertainty

The interviewed end-of-life care providers seemed to take for granted that ethno-cultural sameness as opposed to ethno-cultural difference lies at the core of how high quality and user-friendly care can be provided. They seemed to expect patients with migrant backgrounds to be different from what they perceived to be the norm (which in this case was 'Swedish patients'). As such, caring for patients with migrant backgrounds was described as something that entailed dealing with the unknown:

> You never know what to expect. I have been searching for some information about it, we've asked about it, if we could have some course and get some more information about this ... what we can expect when somebody belongs to this or that religion or this and that ... but nothing has happened. We have a brochure that we found somewhere but it's more about the moment of death and afterwards and how to take care of the dead / ... / but yes there is a lot that we think is strange because it is so foreign to us. (Focus group #1)

In this extract we see two interesting things. First the end-of-life care provider quoted here takes for granted that dealing with patients with migrant backgrounds is bound to be challenging because one does not know what to expect. Something else worth noting is that the unexpected is regarded as 'strange' and 'foreign to us' and therefore requires deciphering. This expectation seems to awaken the feeling that one needs information or education in order to be prepared to handle cross-cultural interaction in this setting. The idea is that dealing with patients with migrant backgrounds entails encountering what some of the informants in other interviews referred to as 'the unknown', 'the unexpected' and/ or interacting with 'a culture that we do not understand and that we are not used to'.

Interacting with the unknown and expecting difference from the get-go were also described as things that awakened a certain degree of fear. This was discussed in one of the focus groups in the following way:

> I think there is a pretty big fear about these issues ... that one might not live up to the patient's expectations ... and just what you guys were saying about the cultural and the religious ... whether or not one has sufficient knowledge to meet this ... I think it feels, you become a little like ... oh my God, how is this going to work?
> / ... /
> / ... /
> These are religions that one isn't involved with so there is of course fear. In Sweden, we have our Christianity and one is very submerged in that but if you come across another religion, it is not so ... you do not really know the rituals so you become a bit insecure, I think. That is how I feel. (Focus group #6)

In this extract we see not only the expectation that the unknown lurks in the background but also that this expectation makes one feel 'fearful' and 'worried' that one will not be able to 'live up to the patients' expectations'. It is worth noting that although both of these data extracts referred to the expectation of difference as something associated with cross-cultural interaction across religious borders, the focus groups discussed the unknown with respect to a variety of things: from not knowing how to behave and worrying about making unnecessary mistakes as a result of one's ignorance, to not knowing how to communicate with and/or interpret what one faces when language barriers exist. The focus on religious difference was, in other words, not always central in the discussions even though it is the angle that this particular data extract focuses on. It is also worth noting that the interviewees made assumptions not only about patients with migrant backgrounds but also about patients without them (which in this case means 'Swedish patients'). We suggest therefore that the expectation of difference that underlies the end-of-life care providers' discussions about what makes cross-cultural interaction challenging gives us insight into why caring for patients with migrant backgrounds is regarded as caring that cannot be offered in a 'business as usual' manner and why caring for patients considered to be 'different' can make one insecure.

One of the numerous ways in which the interviewed end-of-life care providers talked about their expectations of difference had to do with what they expected migrant families to be like. In all focus groups, the idea that these families were fundamentally different from 'Swedish families' was discussed, and thereby, stereotypical ideas about them as well as Swedish families were voiced. The following interview extract is a poignant example of this:

> It is almost a prejudice if one goes to a family whose surname does not sound Swedish. Then one thinks OK, now I have to find out some things and maybe they are in fact the same as me/ … / Because when you go to someone who is … whose surname is Svensson then it is quite a high probability that one will encounter a trodden path because you have done this before. If you go to someone named something else, there are some paths to choose which one must test in order to get where you are going. Maybe you have to go back and try the next path or it is the path you are on that is trodden. So there is more testing somehow. (Focus group #8)

Thus, although the idea that meeting the unknown and expecting difference underlies this data extract as well, it is the implicit allusion to migrant families being different from Swedish families – different from the kind of families we are used to and know how to interact with and/or what to expect when dealing with them – that we draw attention to here. We suggest that the interviewees' expectation of ethno-cultural difference feeds off their assumption that ethno-cultural sameness – having the same background as one's patients – makes interaction in care settings much easier.

Expecting the Extension of Responsibility = Worrying about Facing Dilemmas

When we asked the end-of-life care providers interviewed what they associated with patients with migrant backgrounds, we found much consensus on the issues that they spontaneously brought attention to. The first ideas that seemed to come to mind as far as migrant families were concerned were ideas that were verbalised through short utterances which were seldom questioned during the focus group interviews. These are some examples: 'a lot of people in the ward, a lot of relatives', 'they often have bigger families', 'many relatives around, many to take care of at once, not just the patient'.

The interviewees talked, in other words, about interaction with patients with migrant backgrounds as interaction that was different, not only because one had to deal with people whose backgrounds were different to one's own but also because this interaction entailed, among other things, interacting with families that were larger in size. In this respect it seems worth noting that the sheer size of these families was described as something that could pose problems because, as the interviewees phrased it: (1) 'it can get noisy'; (2) 'in the end there is no air if there are ten people in a room at the same time'; (3) bigger crowds often mean more people who 'disturb the other patients who are there and who do not have the strength to cope with being surrounded by ten people who are talking at the same time'; (4) if 'there are always relatives there (in the room) then one seldom gets to be alone with the patient and cannot therefore establish a relationship with them' and because (5) 'saying farewell takes longer when more people are involved and because the sheer size of migrant families means that there are more people to communicate with and therefore more time to be spent per patient/family'. Thus, migrant families were assumed to create an array of dilemmas which ranged from those posed by lack of space to those resulting from being hindered in creating a relationship with the patient. All of these things meant that caring for patients with migrant backgrounds was expected to entail interaction that stretched the boundaries of responsibility that the interviewed end-of-life care providers were used to.

In the following extract we see how the idea of migrant families as families that are larger and/or that cannot pass unnoticed and that therefore demand different things (such as, more space) underlies some of the discussions that the care providers had as they verbalised what they thought to be challenging about cross-cultural interaction:

> There has also been a need for suitable premises, if I have understood things correctly, people have needed a place to be / … / there are many (relatives) who come from other countries, they might have travelled from really far away and they expect to mourn for several days and, of course, one respects their faith and what they want to do but it is so sad that we do not have the resources or opportunities or the facilities, or whatever it may be that they need.
> / … ./
> The number of relatives who come to say goodbye may differ, there are more. And the process becomes longer in that they may also take longer when they say goodbye.
> / … /
> And large families are a bit more demanding if one is to generalize/ … / One must give a little more time in those cases … (Focus group #1)

In this extract we see how the end-of-life care providers discussed the numerous challenges that bigger families can entail because they obviously require more space, and care facilities cannot always accommodate them. Not being able to accommodate these families was something that posed a dilemma since one had to decide which patients or families one would focus on; the more demanding families of migrant patients or the families of non-migrant patients who might need quieter surroundings? Another issue which was discussed in terms of expectations of diffused responsibility had to do with the fact that migrant families were assumed to be much more hands on as far as caring involvement was concerned. In some instances this high level of involvement was described as something that was positive for the patient (who had company and could feel safe in such a vulnerable situation) and for the staff (who had an extended pair of eyes and ears they could count on). There was, however, a down side to this involvement, as expressed in the following interview extracts:

> They don't let us do what we can do at times because they believe that it is not good for the patient. For example, with medications, and things like that, or whatever it may be ... They want to control things themselves. (Focus group #2)
> Often, they want to care for their loved ones in a completely different way than we Swedes do or maybe not only Swedes but from other Nordic countries. They are involved in the entire process in a different way. We may not get to actually care for the patient ourselves, it is the relatives that are expected to do so and it can be a little difficult at times because we are, after all, are responsible for the care. One ends up in a bit of a dilemma in that case. (Focus group #11)

In these extracts we see not only that these families were always juxtaposed to 'the norm' (which in this case was 'Swedish families') but also that the extended networks of affiliations believed to be typical of patients with migrant backgrounds blur the division of labour between the staff and their families. In addition, when dealing with these patients and their families new areas of responsibility arise, such as those explained earlier: the need to arrange for larger facilities; the need to exchange information with more people; the need to handle more demanding families and/or families with different expectations; and the fact that dealing with more people could be time-consuming. This is why the end-of-life care providers interviewed tended to refer to them as a challenge. Thus, interacting with patients with migrant backgrounds entails handling expectations that extend the realm of responsibility that the providers seemed to associate with 'normal' interactions in end-of-life care settings. This expectation also meant that the interviewees worried about having to face different dilemmas when dealing with these families. It was as if dealing with patients with migrant backgrounds and their families meant dealing with an extended 'patienthood'.

Expecting Misunderstandings = Worrying about Unmet Needs

The most obvious allusion to the expectation described in this section's title had to do with the misunderstandings that lack of a common language can create. During the focus groups, we asked the end-of-life care providers interviewed an open question about what they associated with patients with migrant backgrounds. Remarks that were often made when we posed this question were along the lines of: 'language difficulties, I think ... it is not always that one has access to an interpreter', 'not being able to ask them or not being able to have them tell us what they want because of the language barrier', 'it is not always easy because often it is language that is the biggest problem', 'communication challenges' or 'language, I feel that that is the first thing I think of ... do they speak good Swedish? Can we make ourselves understood to one another?' Thus, expecting misunderstandings brings about uncertainty because patients with migrant backgrounds are assumed to have difficulties understanding the health-care system in general or the various services that are often offered in end-of care units in particular. The following is an example of this:

> When we are dealing with language difficulties, it's important to really ascertain whether or not they have understood the information / ... / if you think about immigrants who have not been here that long ... if you think of Swedes ... they have some knowledge of the system and

> how funeral services work and things like this ... but with them you may need to provide more information and be clearer /../ One cannot assume that they know as much as / ... / It is even more important to be prepared, then ... .to have talked about things before. No, we cannot take for granted that they know. This thing with rehab, for example, which means you can access the services of an occupational therapist and physiotherapist and different types of aids / ... / it seems like it is not obvious that one can get things like that from the society / ... / So it all boils down to making sure that one does not forget any of the things one needs to inform patients about / ... / and if there are language difficulties one needs to make sure that they understand what we mean. (Focus group #5)

The focus group discussions revealed the expectation that one could miscommunicate with patients with migrant backgrounds not only because of the anticipated language barriers but also because these patients lack understanding of what can be expected of end-of-life care. This means that the interviewees anticipated being unable to meet these patients' needs. Another reason why they seemed to expect misunderstandings when dealing with patients with migrant backgrounds was that they believed that the end-of-life care setting was associated with the need to verbalise the anxiety that having one's life come to an end can bring. In the following extract we see an example of this:

> It is a challenge if they cannot speak Swedish. It's very difficult to interpret then. And then also, to be able to convey to this critically ill patient who may be in the dying stages ... to be able to help them calm down and talk to them in the way we do in normal cases, all of these little things that you say when you are in the room with them ... you can't say those things because they cannot understand you/ ... / So I think that it is actually difficult ... And we do not have an interpreter in these situations. If there is someone who is dying, there is no interpreter. (Focus group #9)

In this extract we see that the interviewed end-of-life care providers took for granted that cross-cultural interaction in this setting entailed communication challenges because lack of a common language meant that they could not rely on their ability to soothe the patients as they face death. The discussion in one of the focus groups poignantly described this:

> These patients may not be able to talk about feelings with us in the same way as Swedes do because they do not know the language.
> We end up with sign language / ... /
> Yes, and it is, after all, quite dry many times. Sure, I mean, do you want food? Yes food but ...
> / ... /
> You cannot have a conversation about existential issues, for example, with a human being you cannot communicate with. (Focus group #4)

Thus, although one could sometimes resort to sign language, could get by with short question and answer sessions, and could use one's body in a charade-like manner (which were all things the interviewed care providers expected they would have to resort to when caring for patients with these backgrounds), the end-of-life care setting was regarded by the care providers interviewed as a setting for meaningful conversations. Not being able to communicate with a dying patient about their fear of dying, their existential ruminations and all of the unfinished business that the last stages of life tend to draw attention to was something that the interviewed care providers worried could lead to unmet needs.

Another angle from which expectations of misunderstandings were discussed had to do with the taken-for-granted assumption that patients with migrant backgrounds do not want to know that they are dying. In the following extract we tap into a discussion about a workshop on different communities that some of the interviewed care providers in this particular focus group had attended:

> We have had one of those and it is among other things there that this came up, that one should not, that one should withhold information about the diagnosis to these patients. One protects them in some way / … / we were at a palliative care forum and there were representatives from a lot of different cultures and religions … it was a small forum where they got to inform us … this is how we think and this is how we feel, and it felt very strange to sit there and listen to them … is it really what they think? / … /
> But what was it that they said?
> It was just that … that it is a protection we have, they should not have to go through this, we take over the suffering in some way, my dad shouldn't have to feel this and know this … he should be able to pass away without really knowing what is happening. And it was a bit confrontational, I mean it is bound to be that way since here we have people that have to work according to the health care laws and it is our duty to the patient … .he is supposed to know what is happening. So there are huge contradictions. (Focus group #1)

In this extract we get a glimpse of one of the issues that was extensively discussed in the focus groups, namely the expectation that migrant families want to withhold information about the fact that the patient is dying. This was discussed as a dilemma since this expectation meant anticipating having to break the code of conduct for the caring professions in Sweden. Swedish health-care law stipulates that patients have the right to full disclosure about their medical situation and should be given an opportunity to state their opinions regarding the treatments that are offered to them. One of the interviewed end-of-life care providers expressed this in the following manner:

> When the patient knows that he does not have much time left, he can say, well I want to do this and this and this. But a patient that doesn't know does not have that chance. (Focus group #1)

When one anticipates that dealing with patients with migrant backgrounds will entail dealing with families that want to withhold information from their dying relatives, one has to cope with the fact that full disclosure is not wanted, and also that as a result of having withheld information, the patient is unable to make conscious decisions about their medical situation or to prepare themselves for the inevitability of death that is just around the corner. Thus, expecting misunderstandings and worrying about what this could mean in terms of unmet needs was one of the issues that the end-of-life care providers interviewed associated with caring for migrant patients.

Discussion

The project this article is based on focuses on how end-of-life care providers understand cross-cultural interaction. In this article we draw attention to the assumptions they make about migrant patients as they articulate the challenges they expect to face when providing care across cultural, ethnic, religious and language boundaries. It is worth noting that we purposefully use the term understandings as opposed to experiences since most of the end-of life care providers interviewed seemed to have clear ideas as to what cross-cultural interaction entails and who migrants are even though they had relatively little experience of this type of interaction and/or of this group of patients (see the sample characteristics in the methods section).

The analysis shows that one of the reasons they expected this interaction to be challenging – which is in line with previous research (see Kai *et al.* 2007) – seems to be that at the core of their understandings of patients with migrant backgrounds lay expectations of difference. It is these expectations that this article has focused on. As such, the article draws attention to the uncharted territory that cross-cultural interaction is expected to be, the feeling of uncertainty that having to face difference seems to entail, and the expectation that worrying about facing dilemmas and not being able to meet these patients' needs are inevitable when dealing with patients with migrant backgrounds. The expectation that dealing with these patients could sometimes entail the extension of responsibility meant also that the interviewees seemed to feel forced to expand their professional repertoire as they were faced with what we have hereby referred to as extended 'patienthood'. It is worth noting that although they seemed to be willing to accept this, they worried about not having the resources, mandate, time or competence to do so. In addition, the analysis revealed that expectations of misunderstandings were at the very core of how the interviewed end-of-life care providers regarded patients with migrant backgrounds. This meant not only that the provision of care was regarded as challenging, but also that they envisioned facing a variety of dilemmas because meeting the needs of these patients could, in some cases, mean having to bend some of the rules to which they had to conform as health-care professionals. It is perhaps because of all these expectations that the end-of-life care providers interviewed seemed to dread cross-cultural interaction.

Although the interviewees seldom spoke about patient-centred or culture-competent care per se, we suggest that assumptions about their inability to provide these types of end-of-life care are at the core of the understandings of cross-cultural interaction that the analysis sheds light into since they seemed to take for granted that such care requires that one shares a common background with the patient, has similar frames of reference and speaks the same language. This raises numerous questions if juxtaposed against the literature on end-of-life care which claims that 'culture-competence has evolved from making assumptions about patients on the basis of their background to the implementation of the principles of patient-centred care' (Epner and Baile 2012: iii34). The end-of-life care providers interviewed in this project took for granted that they were unable to provide these types of care because ethno-cultural sameness as opposed to ethno-cultural differentness is needed if one is to provide end-of-life care that is patient-centred and culture-competent.

The focus on difference that lay at the core of the interviewed end-of-life care providers' understandings of patients with migrant backgrounds can be understood against the backdrop that the notions of 'Otherness' and 'Othering' offer. According to Pickering (2001) – who has disentangled the theoretical roots of our conceptualisations of stereotyping and 'Othering' – 'conceptions of the Other and the structures of differences and similarity which they mobilize do not exist in any natural form at all/ … / the location of the Other is primarily in language' (Pickering 2001: 72). 'Othering' is, in other words, the name of the process used to denote how boundaries between the Self and Others are created through categorisation. It is through 'Othering' that we identify and describe those thought to be different from 'us'. This is why 'Othering' practices reinforce our understanding of normality and denote difference as deviance (Pickering 2001). Within the health-care context this term has been used to problematise the practices through which health-care professionals create distance between themselves and patients whose ethno-cultural backgrounds they do not share; practices that magnify the differences which they can sometimes take for granted when caring for these patients (see Johnson *et al.* 2004). In this respect it seems worth pointing out that those who engage in 'Othering' practices are seldom aware that they are doing so. This is important to keep in mind because although 'Othering' is believed by some to be an inevitable process, the ways in which we construct 'Others' as different can take numerous forms (see Wimmer 2008). It is, in other words, through the ways we refer to and talk about the 'Others' that their Otherness is constituted.

Our findings raise a question that the literature on patient-centred care does not really address, namely: what do we need to do and how should we work to develop end-of-life care so that we can face the uniqueness, uncertainty and unknown territory that caring for a patient considered to be 'the Other' seems to entail? Phrased differently one could say that our analysis lead us to suggest that care providers who expect sameness to be a given if one is to deliver patient-centred care, are in fact failing to understand the very ontological premise that underlies the debate on patient-centredness. The debate – and the few scholars who have engaged in it from the culture-competent paradigm, such as Epner and Baile (2012) – stresses the uniqueness of *all* patients and assumes that patient-centredness and culture-competent care are two sides of the same coin. The end-of-life care providers we interviewed however assumed that sharing a common ethno-cultural frame of reference was a given if one is to succeed in providing care that is patient-centred, which is why they anticipated so many challenges when dealing with patients they regarded as 'Others'. It is also worth noting that the end-of-life care providers interviewed voiced numerous stereotypical understandings not only about migrant patients and their families but also about patients who belong to the ethnic majority (Swedish patients) in the context in which we collected our data. It is against this backdrop that we suggest that if we are to take seriously the notion of uniqueness that lies at the core of the debate on patient-centredness then we need to understand that every single patient – irrespective of their ethno-cultural background – is a unique individual whose 'Otherness' – even if they resemble the Self – needs to be taken for granted. Thus, the ultimate question this article draws attention to is: can we really provide end-of-life care in a patient-centred and culture-competent way if we do not regard all patients and their families as potential 'Others' whose uniqueness we must make an effort to decipher?

Disclosure statement

No potential conflict of interest was reported by the authors.

Funding

The authors gratefully acknowledge the small grant from the Medical Research Council of South-east Sweden awarded to Anna Milberg (Linköping University) which made possible the data collection as well as the financial support that the Faculty of Social Sciences of Uppsala University awards to Prof Sandra Torres which has made possible all other phases of the project.

References

Balint, M., 1955. The doctor, his patient and the illness. *The Lancet*, 265 (6866), 683–688.
Creswell, J.W., 1998. *Qualitative inquiry and research design: choosing among five traditions*. London: Sage Publications.
Doorenbos, A.Z. and Schim, S.M., 2004. Culture competence in hospice. *American journal of hospice and palliative care*, 21 (1), 28–32.
Dunn, G.P. and Miller, N., 2014. Patient-centering approaches for the surgical oncologist: palliative care, patient navigation, and distress screening. *Journal of surgical oncology*, 10, 621–628.
Epner, D.E. and Baile, W.F., 2012. Patient-centred care: the key to cultural competence. *Annals of oncology*, 23 (suppl. 3), 33–42.
Evans, N., *et al.*, 2012. Cultural competence in end-of-life care: terms, definitions, and conceptual models form the British literature. *Journal of palliative medicine*, 15 (7), 812–820.
Guba, E.G. and Lincoln, Y.S., 1998. Competing paradigms in qualitative research. *In*: N.K. Denzin and Y.S. Lincoln, eds. *The landscape of qualitative research: theories and issues*. Thousand Oaks: Sage, 105–117.
Gunaratnam, Y., 2007. Intercultural palliative care: do we need cultural competence? *International journal of palliative nursing*, 13 (10), 470–477.
Gysels, M., *et al.*, 2012. Culture and end-of-life care: a scoping exercise in seven European countries. *Plos one*, 7 (4), e34188.
Gysels, M., *et al.*, 2013. Diversity in defining end-of-life care: an obstacle or the way forward? *Plos one*, 8 (7), e68002.
Johnson, J.L., *et al.*, 2004. Othering and being othered in the context of health care services. *Health communication*, 16 (2), 253–271.
Johnstone, M. and Kanitsaki, O., 2009. Ethics and advance care planning in a culturally diverse society. *Journal of transcultural nursing*, 20 (4), 405–416.
Jones, K., 2005. Diversities in approach to end-of-life: a view from Britain of the qualitative literature. *Journal of research in nursing*, 10 (4), 431–454.
Kai, J., *et al.*, 2007. Professional uncertainty and disempowerment responding to ethnic diversity in health care: a qualitative study. *Plos medicine*, 4 (11), e323.
Kitson, A., *et al.*, 2013. What are the core elements of patient-centred care? A narrative review and synthesis of the literature from health policy, medicine and nursing. *Journal of advanced nursing*, 69 (1), 4–15.
Lavoie, M., Blondeau, D., and Martineau, I., 2013. The integration of a person-centred approach in palliative care. *Palliative and supportive care*, 11 (6), 453–464.
Molassiotis, A., 2004. Supportive and palliative care for patients from ethnic minorities in Europe: do we suffer from institutional racism? *European journal of oncology nursing*, 8, 290–292.
Morgan, D., 1997. *Focus groups as qualitative research*. London: Sage.
National Board of Health and Welfare. 2006. *Vård i livets slutskede: Socialstyrelsens bedömning av utveckling i landsting och kommuner*. Stockholm: Socialstyrelsen 2006-103-8.
Patton, M.Q., 2002. *Qualitative research and evaluation methods*. 3rd ed. London: Sage.

Perloff, R.M. *et al.*, 2006. Doctor-patient communication, culture competence and minority health: theoretical and empirical perspectives. *American behavioral scientist*, 49 (6), 835–852.
Pickering, M., 2001. *Stereotyping: the politics of representation.* New York, NY: Palgrave.
Scholl, I., *et al.*, 2014. An integrative model of patient-centeredness: a systematic review and concept analysis. *Plos one*, 9 (9), e107828. doi:10.1371/journal.pone.0107828
Selman, L., *et al.*, 2014. Holistic models for end-of-life care: establishing the place of culture. *Progress in palliative care*, 22 (2), 80–87.
Statistics in Sweden. 2013. http://www.scb.se/en_/Finding-statistics/Statistics-by-subject-area/Population/Population-composition/Population-statistics/Aktuell-Pong/25795/Yearly-statistics–The-whole-country/26046/visited 12/08/2015.
United Nations. 2013. *International migration report 2013.* New York: United Nations, Department of Economic and Social Affairs, Population Division.
Warnes, A.M., *et al.*, 2004. The diversity and welfare of older immigrants in Europe. *Ageing and society*, 24 (3), 307–326.
Wiking, E., Johansson, S.E., and Sundqvist, J., 2004. Ethnicity, acculturation, and self-reported health: a population-based study among immigrants from Poland, Turkey and Iran in Sweden. *Journal of epidemiological community health*, 58, 574–582.
Wimmer, A., 2008. Elementary strategies of ethnic boundary making. *Ethnic and racial studies*, 31 (6), 1025–1055.

The Art of Enduring Contradictory Goals: Challenges in the Institutional Co-construction of a 'good death'

Eva Soom Ammann, Corina Salis Gross and Gabriela Rauber

Institute of Social Anthropology, University of Bern, Bern, Switzerland

ABSTRACT
This paper focuses on the normative notion of 'good death', its practical relevance as a frame of reference for 'death work' procedures in institutional elder care in Switzerland and the ways in which it may be challenged within migrant 'dying trajectories'. In contemporary palliative care, the concept of 'good death' focuses on the ideal of an autonomous dying person, cared for under a specialised biomedical authority. Transferred to the nursing home context, characterised by long-term basic care for the very old under conditions of scarce resources, the notion of 'good death' is broken down into ready-to-use, pragmatic elements of daily routines. At the same time, nursing homes are increasingly confronted with socially and culturally diversified populations. Based on ethnographic findings, we give insight into current practices of institutional 'death work' and tensions arising between contradicting notions of a 'good death', by referring to decision-making, life-prolonging measures, notions on food/feeding and the administration of sedative painkillers.

'Death work' is part of daily business in nursing homes, and nursing home staff establish routines of professional care for the dying elderly. From an institutional perspective, these routines of 'doing death' must comply with organisational constraints and at the same time allow for a dignified individual death. Focusing on institutional elder care, this paper explores the dimensions of 'doing death' when the elderly involved are of migrant background. It is argued that the way institutions 'do death' is tailored to a specific conception of a 'good death', which may be challenged by migrant dying. Drawing on classical social interactionist studies on dying (Sudnow 1967, Glaser and Strauss 1968), we will focus on the co-construction of 'good death' in interactions between residents of migrant background and professional care workers (often of migrant background themselves), and on processes of 'doing diversity' while 'doing death'. The paper is based on the research project 'Doing Death and Doing Diversity in Swiss Nursing Homes' funded by the Swiss National Science Foundation as part of a broad National Research Programme on 'End-of-Life' (NRP67, grant number 406740_139365/1). The data presented show that contemporary 'death work' in Swiss nursing homes is constituted by implicit notions of 'good death', which may collide in concrete interactions around migrant 'dying trajectories', the latter being understood in this paper as a broad and open focus on processes of dying, not in the sense of a tool to predict death and to guide professional practice (Glaser and Strauss 1968).

Notions of 'good death'

The concept of 'good death' builds on a well-documented tradition of sociological, anthropological and historical analyses of dying in modern societies (see, for example, Ariès 1981, Hart, Sainsbury and Short 1998, Kellehear 2007, Green 2008, Hahn and Hoffmann 2009). What a distinctive society regards as a 'good death' is in this strand of literature generally linked to societal modernisation and individualisation processes. While the way societies dealt with dying in classic modernity was characterised by a discourse of denial and displacement (Ariès 1981; for a critical overview, see Hahn and Hoffmann 2009, p. 128ff), societies in conditions of 'reflexive modernity' are confronted with a new visibility of death and the philosophical reflections that accompany it (Bonss and Lau 2011). Dying is increasingly individualised (McNamara 2004, Seale and van der Geest 2004) and the ideal of 'good death' is increasingly subject to public debate (Green 2008). What is at stake nowadays, especially in contexts of old-age dying and dying of degenerative diseases, is the question of keeping control by shaping and 'timing' death (Kellehear 2007), for instance by suicide or euthanasia (Kruse 2007, p. 166ff).

A characteristic feature of dying in contemporary Western societies is its delegation from the home/the family/the community to professional institutions: dying and death are 'outsourced' to hospitals, hospices and nursing homes (Hahn and Hoffmann 2009). While death in nursing homes has always been a part of daily routine which is rarely reflected upon, hospitals became highly specialised, almost omnipotent institutions of restoring and prolonging life, in which death is connoted with institutional failure (Sudnow 1967, Glaser and Strauss 1968). Consequently, the hospice movement emerged as an initiative to reclaim some space for dying and grieving. Hospice discourse has taken over the analytical concept of 'good death' and has transformed it into a normative concept, which is currently entering policies and public discourse. There is increasing awareness also within mainstream health-care services that health care is not only curative (i.e. healing) but in many cases also palliative (i.e. soothing). In consequence, the concept of palliative care (in contrast to curative care) is currently emerging in the Swiss health-care system, not least within biomedical discipline, and services specialising in palliative care are developing both within and parallel to hospitals.

The hospice movement and its reception within mainstream medicine have brought about a rather distinctive normative notion of what a 'good death' is: namely, an individualised and self-determined dying in dignity, peacefulness, preparedness, awareness, adjustment and acceptance (Hart *et al.* 1998). The biomedical prolongation of life at all costs is giving way to the ideal of stopping life-supporting treatments at the end-of-life and instead focusing on pain relief and quality of life. Referring to the Swiss National Strategy on Palliative Care (FOPH and CMH 2012) as an example, palliative care is characterised by an orientation towards the values of self-determination, dignity as well as acceptance of illness and dying. In accordance with the World Health Organisation's definition of palliative care, the Swiss strategy also highlights that caring for the dying should not only soothe physical pain, but also address psychosocial and spiritual needs. Furthermore, palliative care is inspired by the idea that all members of society should have equal access to it.

This conception of dying a 'good death' is, however, tailored to autonomous patients suffering from malignant diseases such as cancer, where a death prognosis is relatively clear and a certain time of preparation for death may be expected (see Kellehear 2007). The capacity to be aware and self-determined furthermore needs a considerable amount of resources, mainly knowledge and communication skills (Schneider and Stadelbacher 2012) but also economic and social resources. The above-described contemporary notion of a 'good death' is an ideal shared in particular (secular, individualistic, Anglo-Saxon) societies (Walter 2012) and mainly compatible with well-educated, middle-class, middle-aged individuals. As will be shown below, this implicit frame of reference is not necessarily shared by all the actors involved in 'dying trajectories', especially – however not only – when migrants are involved.

Negotiations of diversity and migration

As a recent literature review of migration and palliative care (Soom Ammann and Salis Gross 2014) has shown, there is a considerable body of academic writing about palliative care-giving when faced with dying patients of migrant background and their relatives. The literature clearly shows that needs, expectations and specific understandings of central concepts may collide at the end-of-life (see Spruyt 1999, de Graaff *et al.* 2010). What is a 'good death' for the Western biomedical professional may therefore be different from what the patient, their relatives or their community regard as being a 'good death'. Conflicts arise in the absence of an adequate negotiation of differing needs and expectations between all the actors involved in a 'dying trajectory'. Moreover, tensions may also arise due to unsuccessful negotiation processes. Professionals may for example be irritated by relatives' refusal to allow medical staff to communicate openly with the patient (for example, concerning a bad diagnosis or prognosis). Equally difficult from a professional's viewpoint is the refusal to engage in active decision-making (for example, on whether or not to stop treatment) and the insistence on life-prolonging measures or decisions perceived as not compatible with professional ethics (Soom Ammann and Salis Gross 2014).

Underlying these conflicts are diverse cultural, religious or ideological reference systems of what a 'good death' should be like, as the literature suggests and as we will show in detail in this paper. Such frames of reference shape interactions on an implicit level (see 'implicit knowledge' in Giddens 1984). Transforming them into explicit knowledge as a basis for mutual negotiation and consensual decision-making is in practice a very demanding task involving the risk of threatening 'ontological security' (Giddens 1991, Gunaratnam 2008). Specifically, although migrant dying may show very obviously how divergent notions of 'good death' can collide, these conflicts do not only occur when migrant dying is at stake, and they do not only occur between 'death workers' and relatives, but also among relatives and/or among 'death workers'.

'Good death' in institutional aged care

Dying in contemporary modern societies is characterised by longevity: 'natural deaths' at very old age (in opposition to infectious, accidental or violent deaths) are increasingly common. But instead of being the idealised (for instance, controllable and arranged) 'good death', in practice it is often a slow and socially isolated dying process in institutional settings of specialised organisations (Hoffmann 2011), characterised by 'shameful' loss of autonomy and dignity (Kellehear 2007). Therefore, end-of-life in very old age is on the one hand becoming common, but on the other hand, often does not fulfil the ideals of individualised, self-defined and self-controlled dying.

In late-modern societies, caring for elderly in nursing homes is a well-established long-term care option (on Switzerland, see Hoepflinger *et al.* 2011). Due to the general trend in elder care to maintain individual autonomy as long as possible and to transfer persons to institutional care as a last option, nursing homes are increasingly becoming institutions of dying, comprised of residents at very old ages, with advanced multi-morbidity and short duration of stays. In Switzerland, institutions for long-term care are now the most common place of death for the population aged 80 and more (Hoepflinger *et al.* 2011, p. 100). Yet, both research and organisational initiatives still tend to focus on the hospital and specialised palliative care services (see for example, Pleschberger 2007). Furthermore, while in the context of public debates on 'dying well' there are increasing expectations on nursing homes to provide individualised high-quality care, the old-age long-term care sector is under pressure to save costs and is highly affected by shortages of qualified staff. Long-term care institutions are therefore challenged by efficiently providing 'good deaths' to an increasingly pluralised population of elders with a highly diversified staff regarding qualifications, professional careers and national origins.

Issues of 'doing death' while at the same time 'doing diversity' (on the concept of 'doing', see West and Zimmerman 1987) are therefore expected to be at stake in nursing homes, but resources to practice 'good deaths' in this context are limited. The research project from which our data are drawn explored the current practices of dealing with dying and diversity in two Swiss nursing homes that offer standard services for elderly residents in need of care. Both nursing homes are situated in an urban environment within neighbourhoods characterised by heterogeneous populations, including migrants. The two homes care for approximately 120–130 residents each. Nursing home A provides single rooms and is organised in households, one of them offering migrant specific services to 'Mediterranean' residents. These residents mostly migrated as 'guest-workers' from Italy several decades ago. Nursing home B provides mainly double rooms and is organised in wards without specific specialisations in care. In both organisations, the first and second authors conducted extensive ethnographic fieldwork (nine months of participant observation and interviewing) to understand how end-of-life in nursing homes is constituted, how dying and death are 'done' through daily interactions and how diversity issues are at stake in this 'doing death'. Our research methodology followed the principles of Grounded Theory: we entered the research field with a very open research question, data were collected through observation and informal talks with carers, residents, relatives and other actors present in the field, and recorded in extensive field notes. Interviews were selectively audio-recorded (if circumstances allowed for it) and transcribed. The material was edited and analysed in German; for the purpose of this article, selected data were translated into English by the authors. Data analysis followed the classic, theory-generating principles of Grounded Theory assisted by ATLAS TI software. Following the ethical standards formulated by the Swiss Ethnological Society, approval was granted by the management of the two nursing homes. Research participants were initially informed orally on the aim of the research project, and their willingness to participate was constantly re-evaluated by the researchers in concrete interactions.

For the purpose of this article, our selection of data and analytical focus centres on residents with migrant background. While in other papers we elaborate migrant and ethnic diversities in more detail and not only among residents but also with respect to staff and field researchers (see van Holten and Soom Ammann 2016), in this paper we focus on the staff's role as professional carers in dying trajectories, meaning, their 'death work'. This is also in line with our observations in the field that carers tended to strongly focus on their professional roles in their 'death work', and that they treated the two field researchers (first and second authors, both native Swiss) primarily as staff members, despite the authors' repeated and at times also rather explicit enactment of their researcher roles during participant observation in the field.

'Death work'

In approaching 'death work' and its professional procedures of everyday 'doing death' in the nursing home, we refer to the classic study of hospital 'death work' by Sudnow (1967). In this context, 'death work' refers to professionalised routines of those who are occupationally involved in 'doing death'. In the nursing home context, this is mainly referring to care workers of a variety of qualifications (nurses and nursing aides), but also to other occupational actors such as doctors, specialised nurses, social workers, spiritual specialists, cleaning and housekeeping personnel. The main duty of 'death work' involved in Swiss nursing home 'dying trajectories', however, remains with the nurses and nursing aides working on regular day or night shifts. In focusing on 'death work' in this paper, we emphasise 'doing death' from an organisational perspective of professional practice, within an institution defining more or less strict social rules to guarantee its functioning and the fulfilment of its tasks (see for example, the 'total institution', Goffman 1961). From this perspective, the everyday 'death work' of the institution's professional care workers is to be understood as a form of agency with considerable scope for flexibility, but nevertheless obliged to fulfil the requirements of the institution. Therefore, 'death work' tends to focus on routine procedures and avoidance of disruptions to routine (Sudnow 1967, p. 169).

Our analysis has shown that the normative notion of 'good death', inspired by hospice ideology (Hart *et al.* 1998) and in practice adapted to a 'good enough death' within the institutional logics of professional palliative care (McNamara 2004), has also diffused into the nursing home, but in a very practical and intuitive manner. While nursing homes for several decades were oriented towards curative medical practice, they are now starting to shift their orientations towards ideals of palliative care (Kostrzewa and Gerhard 2010). Both orientations are present in everyday practice and shape negotiations about concrete care actions among staff and between staff and relatives, especially in the last days and weeks of life. Residents of nursing homes, however, are often not explicitly involved in decision-making and in negotiating care. On the one hand, this is due to the fact that residents increasingly enter the nursing home with severe cognitive restrictions (such as dementia). On the other hand, speaking openly about death and dying remains a taboo for this generation, especially in the vulnerable moment of having to enter a nursing home (Salis Gross 2001). It is furthermore still very uncommon that residents have drafted 'living wills' to state their wishes for end-of-life care, even though promoting such advance directives is presently one of the main instruments of policymakers to publicly enhance self-determination in dying. Compared to other dying institutions, the nursing home is thus confronted with a rather specific form of dying:

> In most cases you can accept death. It is normal, part of daily life in the nursing home. The residents have had their lives and it is natural that it comes to an end. (Nurse, Daria, Organisation B)

In this sense, dying at a very old age in a nursing home may be seen as a 'good death' since it is at the right time, after having lived a long life. 'To be able to go' or 'to let go' are standard expressions in the nursing home to talk about dying, without having to be too explicit about it. These two notions furthermore point to the dominant conception of dying understood as an individual, self-determined action. What Daria describes above is the ideal of a 'natural' death where the resident decides when it is time to die and the carer only has to make sure that nothing is hindering the passage. 'To allow it' or 'to let it happen' are common terms to talk about the role of the carers. In practice, this is most often interpreted as making sure that the resident is not disturbed unnecessarily, that he/she is lying comfortably and does not feel dehydrated. To ease physical suffering and anxieties, morphine is widely used by the 'death workers'.

There are, however, also 'bad' or even 'terrible dying trajectories' that arise unexpectedly, are not controllable or seem to be unclear, necessitating decisions to be taken. Such decisions lie heavily on the shoulders of the care staff and always contain the risk of taking the wrong decision, as nurse Marina describes when talking about the sudden, indistinct deterioration of a resident:

> I was alone that day, had to bear the whole responsibility. I called her daughter, and she said: Well, what do you think, is it really an emergency? She did not want to decide. [...] Then I called the doctor, told him how it was, and he said: 'Well, if she stays with us in this condition, she's going to die. What do you think should be done?' [...] I told him loud and clear that I am not in a position to decide about life and death! This is something the relatives have to decide, maybe together with the doctor, but not me! (Nurse, Marina, Organisation A)

Marina was in this case in an extraordinary situation, but what she articulates here is something constantly present in the caring work of the nursing home, making it implicitly also a constant 'death work' potentially loaded with guilt (Salis Gross 2001). Our data show that the threat of taking wrong decisions and the implicit possibility of being guilty, which is involved in 'bad' or 'terrible deaths', may lead the professional carers to strive for reconstructions of 'good deaths' (see also Simpson 2001) by investing in 'death work' practices regarded as elements of 'good deaths'. Such strategies include paying more attention to residents perceived as 'difficult' or 'suffering' in terminal phases by spending more time in their rooms, holding their hands or taking over an advocacy role for them. Another strategy to transform 'bad deaths' into 'good deaths' is telling stories about the resident stressing positive aspects of their everyday lives in the nursing home. Furthermore, the preparation of dead bodies seems to be an important occasion to recast 'bad' or 'terrible deaths' as 'good deaths' by investing in preparing deceased residents in an especially 'beautiful' way, trying to give them a bodily expression of peace and dignity, decorating the deathbed with flowers, etc.

> To me it is very important that a dead person looks peaceful, calm, relaxed. We give our best to make the person look nice. It is like a last favour, to make her or him look the way they looked before the suffering started. It is not only for us but also for the relatives. This last picture must be peaceful. It helps a lot to let go. (Nurse, Katha, Organisation B)

Thus, the 'death work' of professional care workers in nursing homes is a constant process of 'doing death' on behalf of the residents. The aim of 'doing' such deaths as 'well' as possible is a shared professional norm for all care workers, but the ideals of palliative care underlying it (as described above) are not known in detail to all of them. Balancing the everyday need to take decisions on behalf of dying residents by interpreting what is to be done to make their individual dying a 'good death' is therefore a demanding task for carers working in nursing homes:

> It's a very complex thing to take decisions in such situations, and it is mostly about insecurities. One feels the need to do something, instead of reflecting on the situation and on the possibilities one has, and here many of us are – yes, they are somehow incapable. It is a demanding task. You need to be able to think in complex terms, to step back, to discuss, to find consensus. Yes! And you certainly are quicker giving an antibiotic or calling an ambulance. (Team leader, Charlotte, Organisation A)

In consequence of the fact that residents often do not or cannot explicitly say how they intend to die before they become terminally ill, decisions about their care in the last weeks and days have to be taken over either by the staff or by the relatives. Although Swiss law gives clear priority to relatives to decide about medical issues, the carers sometimes feel that they are in a better position to decide for the residents since they are the ones sharing everyday life with them, not the relatives. If relatives delegate decisions to the staff or if there are no relatives, the staff members make decisions by reference to their care directives and their personal relationships with a resident (if they for example keep the door open or closed, put the light down, inform other residents, and so on). And, most importantly, they rely heavily on interpreting minimal bodily signs according to their experiences with this resident, to interpret for example if they want to eat or drink, are at ease or in pain. The considerable diversity among staff in terms of their professional training and biographical experiences serves as a frame of reference for individual carers to interpret what is appropriate, as the following case studies will show.

The challenge of 'doing a good death': case studies

To illustrate the points of tensions in providing for a 'good death', four case studies will now be discussed. They focus on practices of adjusting professional 'death work' procedures in order to deal with diversity.

The 'uneased dying trajectory'

Mr S., a German migrant who had lived in Switzerland for 20 years, was a member of a Protestant Free Church. He entered nursing home B at the age of 80. The staff described him as a polite man who led a secluded life (as far as this is possible in a nursing home). He died two years later of lung cancer. In his last days and hours of life he suffered from severe chest pain and had difficulty breathing. Nevertheless, Mr S. continued to verbally reaffirm his will not to receive sedative pain therapy. He wanted to die and 'face God' with a clear mind, as he repeatedly stated. Samantha, the nurse in charge of the nightshift during the very last phase of his life, was extremely challenged by the resident's will. For her, a professional 'death worker' with years of experience in a hospital palliative care unit, standing at the bedside of Mr S. who was agitated, short of breath and 'moaning from pain' proved to be hard to bear. Having the morphine 'always in the pocket' ready to be administered, but explicitly not having the resident's permission to use it, literally rendered her powerless, as she stated. Nevertheless, complying with the legal directive that respect for the individual's will takes precedence, she did not administer any strong analgesia, and the resident died in the early morning hours.

In contrast to most 'dying trajectories' we observed in the nursing homes, Mr S. acted as an autonomous, self-determined individual who clearly expressed his will until the very end of his life. His understanding of a 'good death' led to a 'dying trajectory' that was difficult to accept for the professional 'death worker', whose procedures for 'doing death' focus on easing the resident's pain and thereby making dying 'bearable' for the resident, but also for all the other actors involved. It may be argued that a resident who substantially and consciously shapes his dying himself is to be considered as the ideal patient in palliative care. However, the way Mr S. wished to 'do' his death collided with the nurse's notion of 'good death'. Referring to her own reference system as a professional 'death worker', she felt under pressure to produce a 'good death' in the sense of reducing suffering. She would clearly have preferred to control the situation by alleviating pain instead of accepting the resident's deliberate choice to suffer. Surrendering to his will therefore implied a severe disturbance of her routine as an institutional 'death worker' who is not only taken to her limits by tolerating an opposite view of 'good death', but who at the same time had an obligation to be concerned about collective well-being on the ward. A resident who is moaning from pain during the night also disturbs the other residents, most notably the roommate. This will not only prevent them from sleeping, but also confronts them with their own vulnerable situation. The carer may feel exposed to questions regarding her professional capacity (was not there anything you could do for him?) and guilt (why did not you help him?). Additionally, caring for a dying resident during nightshift, where one nurse is responsible for thirty residents, is already more difficult to organise than during dayshift when more staff are available. In case of a sudden deterioration (such as threat of suffocation), the situation might demand her full attention at a time when other residents should be repositioned, helped to go to the toilet and so on. Residents under sedative and analgesic medication are in this sense easier to control and cause less disruption to the carers' work duties and to the well-being of the other residents. Surrendering to Mr S.'s will, therefore, also meant giving up control over this 'uneased dying trajectory'.

The 'guilt-ridden dying trajectory'

Mrs A., having migrated from Sri Lanka in the 1980s, came to nursing home B after a cancer incident at age 71, being unable to control her legs. She was appreciated as a lovely person by staff and fellow residents, integrating well into the ward, not posing any challenges, and being frequently visited by her husband and son. Her terminal phase began rather unexpectedly a few months after her entry to the nursing home with sudden and severe pain. The nursing staff did not immediately succeed in their attempts to control this heightened pain with adequate morphine doses. At the same time, according to the standard procedure of the nursing home, her relatives were regularly informed about the deterioration of her condition. On the day of her death, husband and son were asked early in the morning to come by. Katharina, the nurse in charge, wanted to tell Mr A. that she thought Mrs A. was actually dying now. Although the son translated this to his father, the nurse was not sure if the latter understood the information she wanted to pass on. In the next hours, the husband stayed with his wife, more and more puzzled that he could no longer talk to her as she was now unconscious. Feeling that she had to justify her procedure, the nurse explained that Mrs A. was feeling 'very sleepy' since she had received morphine and other strong opiates and that she, the nurse, was grateful to see that Mrs A.'s suffering seemed to be eased now. However, Mr A. demanded that the morphine be reduced in order to be able to talk to his wife. Later, he demanded the intervention of a heart–lung machine to keep his wife alive. He mentioned that Mrs A.'s mother had to come and see her daughter, but that she would need another day to travel to Switzerland. The professional 'death worker', considerably irritated by this request, was not willing to acquiesce since she regarded it as an unnecessary prolongation of Mrs A.'s obvious suffering in order to comply with someone else's needs. In accordance with the nursing home doctor's advice and the patient's earlier statement that she did not want any life-prolonging measures, the nurse continued to give high morphine doses. Shortly afterwards Mrs A. died without regaining consciousness. Standing beside the deathbed at that moment, the husband expressed his suspicion that 'they made her go' by giving morphine. Katharina, taking Mr A.'s hand and looking straight into his eyes, replied that it was the cancer that had killed his wife, not the medication. Minutes later, the husband expressed the wish that his wife be dressed in her wedding sari. Katharina, now that the hectic rush of the 'dying trajectory' was over, encouraged him to retrieve the sari while the dead body was washed and prepared for the dress.

Mrs A.'s 'dying trajectory' came on unexpectedly and progressed rapidly, causing visible suffering and thereby putting the professional 'death worker' under heavy pressure to react. Nevertheless, from the perspective of the staff, their procedures of care for the dying resident may be seen as successful and executed according to the nursing home's routine: referring to the palliative care standards, the resident's relief from acute pain and respect for her stated wish not to receive life-prolonging measures constitutes the appropriate frame of reference for professional 'death work'. It therefore could be argued that from the point of view of the professionals and the resident a 'good death' was (at least partly) produced. The collision in this case arises between professionals and relatives. On the one hand, there is not enough time and attentiveness from the professional's side to also consider the husband's conflicting needs. Not only does he express his needs late in the 'dying trajectory', but also rather unexpectedly, thereby upsetting the 'death workers'' characterisation of him as a compliant person who is grateful for every help his wife receives. On the other hand, there is not enough power on the husband's side to insist on his view of a 'good death', namely to be able to say goodbye to those who are important (but who may not be nearby when migration is involved), and to do everything possible not to 'make somebody go'. After the death of Mrs A. and the husband's sudden accusation, Katharina shows considerable tolerance in allowing the husband enough time to retrieve his wife's wedding sari and in organising 'cultural experts', namely staff members of Indian migrant background who, as Katharina supposes, should also know how to wrap a body in a sari. By doing so, the nurse accepts that the professional routine of preparing a dead body quickly and with clothes available nearby is disturbed. It might be argued that she thereby tries to convert a 'guilt-ridden dying' into an experience which also includes elements of a 'good death' from the perspective of the husband.

The 'unprotected dying trajectory'

Mr R., suffering from advanced dementia, was a resident for almost a decade in nursing home B. When ward leader Beata talks about his terminal phase and death, she mainly talks about the relatives. From the beginning, she says, Mr R.'s children were always very emotional about their father, and she experienced them as rather domineering towards the staff. She says that they explicitly questioned on repeated occasions if Mr R. was being adequately cared for since the staff were not in a position to understand his 'Italianness', as they put it. As a manifestation of this point, the relatives used to come by every evening and bring home-made food, which they fed to him. When Mr R.'s condition deteriorated towards the end of his life, he did not talk any more, he was not able to get up, and he stopped eating autonomously. At this point, the relatives increased their presence in the nursing home and, even though Mr. R. did not swallow anymore, they continued trying to feed him with their own food. In Beata's words, they 'prodded him, opened his eyes, stuffed him', they literally 'forced' him to eat, and they told the staff that 'food would keep their papà alive'.

It was, as Beata says, very difficult for the relatives to understand why the staff did not feed their father anymore in the morning and at lunchtime; they thought that the staff were going to let him starve. Opposed to this, Beata would have preferred to stick to the usual procedure of professional 'death work' in the nursing home, whereby not swallowing food autonomously is interpreted as a 'refusal', i.e. an expression of the resident's will that has to be respected by professional 'death workers'. Therefore, the relatives' insistence on eating not only collided with the usual procedures of the nursing home carers, but their 'prodding' and 'stuffing' was experienced by the staff as a veritable assault on Mr R., which would necessitate protection by the professional 'death workers' responsible for his well-being. On the other hand, close relatives are by law the first proxies for patients, and therefore Beata's professional understanding of 'doing good death' also demanded compliance towards relatives' wishes. She therefore tried to balance her advocacy for the resident with the wishes of the relatives to 'keep their papà alive'. However, the relatives' conflicting views of how to properly care for their father, together with their domineering behaviour towards the staff and questioning of their professionalism, were a key structuring element of the relationship between carers and relatives which Beata did not succeed in resolving. Even after Mr R.'s death, the relatives continued to make claims on the staff and expressed 'strange' wishes: they wanted their father to be dressed in a black suit, including shoes and hat, something that bewildered Beata.

Beata's narration of Mr R.'s death is an example of a 'dying trajectory' in which the patient's agency was absent. As is stipulated by law, his children took over as his proxies. What was challenging for the professional 'death worker', however, was that the relatives referred to a vision of 'good death' conflicting with her view. The relatives clearly insisted on doing everything possible to prevent death. Furthermore, they were very active in doing what, as they thought, was within their power to provide for a 'good death': making sure he ate what they cooked for him. Beata clearly disapproved with this interpretation of a 'good death'. She would have stuck to what palliative care standards suggest, namely accepting that the patient was signalling his wish to die and therefore allowing him 'to go'. Beata tried to be tolerant but succeeded neither in convincing the relatives of the usual procedures of 'death work', nor in preventing them from enacting their roles as 'caring relatives' in the nursing home. While she was not able to accept their behaviour in the terminal phase, she did, however, partly surrender to the conflicting expectations of the relatives after death by dressing Mr R. in a hat and shoes, clothing that did not comply with the standard procedures for preparing a corpse in this nursing home. While it might be interpreted that by doing so she was (partly) successful in enabling the relatives to 'do' their father's death as they felt was appropriate, she was obviously left with a bad feeling, leading her to generalise from this experience that Italians as an ethnic group are difficult nursing home customers. This may be interpreted as a strategy to deal with the potential guilt of having failed in protecting the resident from assaults on his right to die in peace by locating the explanation outside of her scope of agency, namely in the supposedly ethnic nature of the relatives' behaviour.

The 'indecisive dying trajectory'

Mrs I. was a severely demented Italian migrant living in the 'Mediterranean' ward of nursing home A for several years. She was rather mobile, but had retreated to only speaking a local dialect nobody else in the ward was able to understand, and she had been eating and drinking badly ever since she entered the nursing home. During a heatwave period in summer, she developed a fever and diarrhoea, and she no longer got up or even reacted to the carers' presence any more. After some discussions between the doctor, the nurse in charge (Marina) and the daughter, it was decided to send Mrs I. to hospital, where dehydration was diagnosed. After a few days, Mrs I. was returned to the nursing home and seemed to have recovered somewhat. Nurse Marina admonished the staff that dehydration in Mrs I.'s case was a failure to have cared properly for her, since she always had to be encouraged to drink. Feeling guilty about Mrs I.'s condition for not having paid enough attention to her drinking, Marina was in a state of increased attentiveness.

Mrs I. then seemed to have more difficulties in swallowing liquids than she had before. The doctor called the daughter to discuss what was to be done. He suggested the option of prescribing a hydrating infusion to give Mrs I. a chance to further recover, but he could not tell if she would. He wanted the daughter to decide, but she refused and instead called for her brother, who was living in Italy. He travelled to the nursing home, but could not decide either. Marina and the doctor saw that he was crying and therefore decided to give Mrs I. an infusion: this also gave her son some time to accept that his mother might be about to die. Mrs I. got the infusion, and nurse Marina eagerly observed her condition during the following days. However, Mrs I. did not show any clear signs if she was recovering or 'only kept alive by the hydrating infusion'. Marina felt that it was necessary to take a decision, but perceived this as 'a decision over life or death', which she was not willing to take since for her there were no unambiguous signs. Referring to the palliative care ideal of respecting the patient's will, she said that she was not able to come to a clear conclusion on Mrs I.'s wish to live or die. Was Mrs I. at all in a position to express her own will due to her advanced dementia? Did she still have quality of life? Was she suffering, or was she at ease? Due to the absence of distinct physical signs, Marina insisted that the relatives decide, or at least clearly express that they would accept it if the staff decided to remove the infusion. It took several conversations with the daughter, the son, the doctor, a nurse colleague and the team leader to come to a decision, and the infusion was removed. It then took another two weeks until Mrs I. died, leaving the relatives with plenty of time to come by, talk to the nurses, prepare for the burial, but also to be a bit puzzled about the ongoing uncertainty of their mother's status, which neither showed signs of recovery nor of suffering even after removal of the infusion.

Mrs I.'s case illustrates that if the dying individual does not or cannot express his or her wishes regarding a 'good death', it is not only challenging for the relatives, but also for the caring staff, who in such situations become more aware of their role as 'death workers'. If the loose, informal guidelines the nursing home has taken over from the palliative care model do not help to come to a more or less clear assessment of the supposed will of the dying person, and if there are no articulated alternative guidelines at hand to which one could stick (such as the relatives' guidelines in case C), it seems to be very challenging to decide what to do when decisions are potentially life-threatening. Someone would have had to take over the duty of 'doing death' on behalf of Mrs I., but no one really did. Furthermore, the action upon which a decision had to be taken – the removal of an infusion – is a very invasive action, a particularly obvious form of 'doing death', which makes it even more difficult to take a decision. Pulling out the infusion is in this case comparable to the action of turning off a life-support machine, having the symbolic meaning of an agency 'deciding about life and death'. Marina struggled with this aspect of professional agency and the inherent pressure in 'death work' to take decisions having potentially severe and irreversible consequences. She questions the hesitant strategies of both doctor and relatives, trying to explicitly negotiate who is in charge of taking decisions on behalf of Mrs. I., while the evidence of what Mrs I. would have regarded as a 'good death' was completely unclear. In Mrs I.'s case, time was on Marina's side, giving her the opportunity to articulate her unease, to critically debate professional roles and powers and to insist on finding solutions mutually agreed upon, thus preventing her from feeling guilty about having taken a potentially wrong decision.

The art of endurance in challenged 'death work'

By presenting four case studies, we intended to illustrate four possible lines of collision regarding what a 'good death' is and how professional procedures of 'doing a good death' may be disturbed in contexts of societal diversities. Focusing on the 'death workers'' perspective, the cases point to different strategies of dealing with disturbed routines. While case A is characterised by the need to be able to acquiesce and endure if a resident insists on a way of 'good death' which is incompatible with professional procedures, the following three cases are all characterised by a dying resident who lacks autonomy as a consequence of the severely constrained ability to show or articulate wishes, a situation which is rather typical in nursing homes. Issues of acting as a proxy on behalf of the 'non-self-determined' resident are at stake in all these cases, and collisions occur between 'death workers' and relatives as well as among 'death workers', as we plan to further elaborate in the future. In these cases, strategies to alleviate collisions are used to maintain or to re-establish professional agency. In case B, the 'death workers' impose their power on the relatives and stick to their routines in the process of dying, but attempt to alleviate their sense of culpability later by handing over agency to the relatives when it comes to the procedures of treating the dead body. Case C is characterised by a constant but in the end unsuccessful negotiation between the 'death workers' and the relatives on what a 'good death' for the resident should be like. In surrendering to the relatives, the professional 'death workers' somehow also fail to protect the resident from what they regard as violating his right to have a 'good death'. Case D shows that 'death work' on behalf of 'non-self-determined' residents is constantly threatened by doing the wrong things and thereby provoking fatal consequences. Here the actors involved implicitly try to avoid making important decisions and in so doing to avoid potential guilt.

'Doing good death' is in all four cases subject to debates since the actors involved refer to 'ontological securities' which are not necessarily congruent. In other words, implicit conceptions of human agency towards issues of life and death may be diverse. Yet, condensing such diversities down to clear 'cultural' categories is not possible due to the complexity and individuality of ontological reference points. Therefore, as Gunaratnam (2008) has shown in detail, reference to 'cultural knowledge' is not in itself helpful in concrete 'death work' (see also Kai *et al.* 2007). The everyday interactions of professional carers', residents' and relatives' 'doing death' are always threatened by insecurities that question routinised procedures: not only with regard to professional 'death workers', as Gunaratnam (2008) has shown, but also with regard to residents and especially relatives acting as proxies. Dealing with 'ontological insecurities' in 'dying trajectories' may lead to a variegation of agency, ranging from explicit agency (as for example the relatives in case C) to incapability of agency (as in case D). The specificity of professional 'death work' in contrast to the agency of residents and relatives is, however, that agency must be maintained or re-gained even in circumstances of severe 'ontological insecurity', in order to fulfil the professional requirements of 'death work' and to enable the carer to move on to the next resident and their particular 'dying trajectory' (see also Kai *et al.* 2007). Therefore, 'deaths' constantly have to be 'done' with reference to past and future deaths, and making them 'good' is never accomplishable without some doubts potentially arising in the minds of the actors involved (see also Salis Gross 2001).

As has been shown, the dilemma described above – between the requirements of the 'total institution' asking for routinised agency and the requirements of 'good death' to provide individualised dying in dignity – is constant and cannot be resolved under the current conditions of modern societies. Furthermore, our case studies show that negotiations about 'difference' in practices of 'doing death' are not a simple matter of clearly definable cultural or migrant backgrounds, but a matter of plurality and individual autonomy in contemporary societies (see also McNamara 2004). Drawing a clear line between diversities within and beyond societies – simply put, between non-migrants and migrants – is a complex thing, and lines of collision do not stick to clear-cut non-migrant or migrant specific reference systems, but become blurred and multifaceted in everyday practice. It is, for example, the 'not so foreign' German resident in case A, belonging to a Protestant Free Church, who challenges the nursing home with rigorous religious argumentation. And while the actors in case C refer to stereotyping ethnic explanations for their diverging ideas on what 'good' care for a dying person is, 'ethnic sameness' on the Mediterranean ward does not facilitate decision-making in case D. While the carer in that case refers to her professionalism to justify not being able to decide, case B illustrates precisely how taking decisions at the end-of-life is associated with guilt. At the same time, case B also demonstrates that 'doing good death' is related to power and resources.

Therefore, 'doing good deaths' in contemporary nursing home 'death work' is about being able to endure the fact that contradictions may arise at any time in 'dying trajectories'. An open awareness of this dilemma, and the provision of instruments to reconstitute agency (or to re-establish 'ontological security') for institutional 'death workers' who are always at risk of being left with feelings of unease or guilt, may be crucial to enable 'good deaths' for increasingly diversified nursing home populations.

Acknowledgements

We would like to thank our colleague Alistair Hunter and the two anonymous reviewers for their valuable comments on earlier versions of this paper.

Disclosure statement

No potential conflict of interest was reported by the authors.

References

Ariès, P., 1981. *The hour of our death.* London: Allen Lane.

Bonss, W. and Lau, C., eds., 2011. *Macht und Herrschaft in der reflexiven Moderne.* Weilerswist-Metternich: Velbrück Wissenschaft.

De Graaff, F.M., *et al*, 2010. 'Palliative care': a contradiction in terms? A qualitative study of cancer patients with a Turkish or Moroccan background, their relatives and care providers. *BMC palliative care*, 9 (19). doi:10.1186/1472-684x-9-19

Federal Office of Public Health (FOPH) and Swiss Conference of the Cantonal Ministers of Public Health (CMH) 2012. *National guidelines for palliative care.* Bern: FOPH.

Giddens, A., 1984. *The constitution of society, outline of the theory of structuration.* Cambridge: Polity Press.

Giddens, A., 1991. *Modernity and self identity.* Cambridge: Polity Press.

Glaser, B.G. and Strauss, A.L., 1968. *Time for dying.* Chicago: Aldinge.

Goffman, E., 1961. *Asylums: essays on the social situation of mental patients and other inmates.* New York: Doubleday Anchor.
Green, J.W., 2008. *Beyond the good death. The anthropology of modern dying.* Philadelphia: University of Pennsylvania Press.
Gunaratnam, Y., 2008. From competence to vulnerability: care ethics, and elders from racialized minorities. *Mortality,* 13, 24–41. doi:10.1080/13576270701782969
Hahn, A. and Hoffmann, M. 2009. Der Tod und das Sterben als soziales Ereignis. *In:* C. Klinger, ed. *Perspektiven des Todes in der modernen Gesellschaft.* Wien: Boehlau, 121–144.
Hart, B., Sainsbury, P., and Short, S., 1998. Whose dying? A sociological critique of the 'good death'. *Mortality,* 3 (1), 65–77. doi:10.1080/713685884
Hoepflinger, F., Bayer-Oglesby, L., and Zumbrunn, A., 2011. *Pflegebeduerftigkeit und Langzeitpflege im Alter. Aktualisierte Szenarien fuer die Schweiz.* Bern: Huber.
Hoffmann, M., 2011. *'Sterben? Am liebsten ploetzlich und unerwartet'. Die Angst vor dem 'sozialen Sterben'.* Wiesbaden: VS Verlag für Sozialwissenschaften.
Kai, J., *et al.*, 2007. Professional uncertainty and disempowerment responding to ethnic diversity in health care: a qualitative study. *PLOS medicine,* 4 (11), e323. doi:10.1371/journal.pmed.0040323
Kellehear, A., 2007. *A social history of dying.* Cambridge: Cambridge University Press.
Kostrzewa, S. and Gerhard, C., 2010. *Hospizliche Altenpflege.* Bern: Huber.
Kruse, A. 2007. *Das letzte Lebensjahr. Zur koerperlichen, psychischen und sozialen Situation des alten Menschen am Ende seines Lebens.* Stuttgart: Kohlhammer.
McNamara, B., 2004. Good enough death: autonomy and choice in Australian palliative care. *Social science & medicine,* 58 (2004), 929–938. doi:10.1016/j.socscimed.2003.10.042
Pleschberger, S., 2007. Dignity and the challenge of dying in nursing homes: the residents' view. *Age and ageing,* 36, 197–202. doi:10.1093/ageing/afl152
Salis Gross, C., 2001. *Der ansteckende Tod. Eine ethnologische Studie zum Sterben im Altersheim.* Frankfurt: Campus.
Schneider, W. and Stadelbacher, S., 2012. Alter und Sterben anders denken – Soziologische Anmerkungen zur Zukunft des Lebensendes. *Die Hospizzeitschrift,* 2012 (3), 6–11.
Seale, C. and van der Geest, S., 2004. Good and bad death: introduction. *Social science & medicine,* 58 (2004), 883–885. doi:10.1016/j.socscimed.2003.10.034
Soom Ammann, E. and Salis Gross, C., 2014. Palliative Care und Migration. Literaturrecherche zum Stand der Forschung einer diversitätssensiblen Palliative Care. *In:* C. Salis, *et al.* eds. *Chancengleiche Palliative Care. Bedarf und Bedürfnisse der Migrationsbevölkerung in der Schweiz.* Muenchen: AVM, 101–200.
Simpson, B., 2001. Making 'bad' deaths 'good': The kinship consequences of posthumous conception. *The journal of the royal anthropological institute,* 7 (1), 1–18. doi:10.1111/1467-9655.00047
Spruyt, O., 1999. Community-based palliative care for Bangladeshi patients in East London. Accounts of bereaved carers. *Palliative medicine,* 13, 119–129.
Sudnow, D., 1967. *Passing On: the social organisation of dying.* Englewood-Cliffs: Prentice-Hall.
van Holten, K. and Soom Ammann, E., 2016. Negotiating the Potato: the challenge of dealing with multiple diversities in elder care. *In*: C. Schweppe and V. Horn, eds. *Transnational aging – current insights and future challenges.* Routledge Series: Research in Transnationalism, 200–216.
Walter, T., 2012. Why different countries manage death differently: a comparative analysis of modern urban societies. *The british journal of sociology,* 63, 123–145. doi:10.1111/j.1468-4446.2011.01396.x
West, C. and Zimmerman, D.H., 1987. Doing gender. *Gender & society. Official publication of sociologists for women in society 1* (1987), 125–151.

End-of-life Care and Beyond

Fuusje de Graaff

Projectbureau MUTANT, The Hague, The Netherlands

ABSTRACT
One of the challenges of palliative care is to honour the personal wishes of culturally diverse patients while meeting 'universal medical relief standards'. Diverging perspectives on good care can result in intercultural negotiations, such as those analysed in this paper between Dutch care professionals and immigrant families with a Turkish and Moroccan background. Fewer tensions are apparent during the burial care phase. This paper outlines the experiences of immigrant families and their Dutch care professionals in the transition from palliative care to burial care. Their narratives highlight how undertakers offer burials that follow the funeral rituals practised in their customers' (predominantly Muslim) home countries and therefore employ bilingual staff to fulfil the specific wishes of the families. Palliative care professionals, on the other hand, tend to follow their Dutch interpretation of 'universal medical relief standards'. The contrast not only reveals cultural diversity, but also different logics of care; while undertakers follow the logic of the market and the logic of family life, health care professionals tend to follow the logics of professionalism and politics.

For centuries care for the dying was a family affair. Yet some tasks were often delegated to social, cultural or religious leaders. In many Western countries end-of-life care has been institutionalised, entrusting some tasks to the hands of professionals. Medical professionals even distinguish the care given at the end of life (palliative care) from curative care, as the focus is no longer on treating a patient but on preserving or improving their quality of life (WHO 2010). The desire to align this care with the specific needs of those requiring it precludes a universal standardisation of this type of care; rather, this will be tailored to the prevailing norms, which can vary per country, region, cultural group, family or individual. For this reason, various studies have recently been conducted in the Netherlands on the specific needs of migrants of Turkish and Moroccan background (Bot 1998, Yerden 2000, 2004, De Graaff and Francke 2003, Yerden and van Koutrike 2007, Buiting *et al.* 2008, De Graaff *et al.* 2010a, 2010b, 2012a). Care providers, patients and families appear to differ on what is constituted by 'good' care and 'appropriate' communication, especially at critical decision-making moments and transition phases. These transitions not only demand culturally sensitive attention regarding physical care, but also with respect to psychosocial, social and spiritual needs.

In this paper I distinguish three transitions in end-of-life care situations that often lead to diverging points of view between migrants of Turkish or Moroccan background and their Dutch care professionals. Focussing on the last transition, I describe some findings from my research and demonstrate that the diverse experiences of migrant background families with palliative care professionals and burial care providers, respectively, can be understood by considering the different logics at work in palliative care and burial care.

Transitions in care situations

Three major transitions in care situations can be distinguished. Although this paper focuses on the third, a short overview of the transitions that come before it is necessary to illuminate some of the problems often said to arise between immigrant families and palliative care professionals in the Netherlands.

The first transition relates to the initiation of palliative care, as an adjunct to or following curative care, due to the patient having been declared to be – medically speaking – 'incurably ill'. In many cases, immigrant patients and their families, in the above mentioned studies, did not allow the curative goal to be relinquished, since, in their view, the course of life is determined by divine powers rather than biological processes (De Graaff *et al.* 2010b, De Graeff *et al.* 2012). As a result, initiating palliative care was not a subject that was readily addressed; relatives were often reluctant to have their loved one told that the end was approaching because they held out hope for a happier outcome. By contrast, Dutch care providers tend to strive for open communication and to provide support to patients facing death (Onwuteaka-Philipsen *et al.* 2007).

The second transition is the point at which care providers wish to refrain from resuscitation, but are considering sedation to treat pain or delirium. Care providers prefer to discharge patients who have reached this stage from the hospital to allow them to die at home. Turkish–Dutch and Moroccan–Dutch patients are more likely to die in hospital (Buiting *et al.* 2008), in part because the families continue to demand curative treatment (VPTZ 2008), and at the same time reject sedation. Some family members of Turkish and Moroccan background stipulated that pain and discomfort may be seen as God's will or punishment and that they wished their loved one to appear before Allah with a clear mind (De Graaff *et al.* 2010b). The fact that they reject sedation and the use of opioids is often presented from a religious point of view, but can also be related to their inability to explain their feelings in a foreign language (Denktaş 2011), and their need to belong to their cultural community (Vroon-Najem 2014). Many Muslim immigrant communities defer to the opinion of their own recognised religious scholars, as the Islamic faith is not monolithic but rather a diversity of views exist (Daar and Khitamy 2001 in Bulow 2008).

The final transition regards death itself, where caring for a person who is ill passes into care for the deceased. Although the majority of Muslims in the Netherlands are anxious to observe the funeral rights and regulations of Islam, the customs can differ from person to person and from family to family (Venhorst 2013). Local Muslim leaders instruct Dutch care professionals that a dying person should preferably be laid on the right side with his or her face tilted towards Mecca, while his fellow believers should gather around to recite from the Qur'an and help the dying person to repeat the Shahada (the profession of faith).

According to the families with a Turkish and Moroccan background, the period between death and burial should be kept as short as possible in order to hasten the transition to the afterlife. The deceased should be covered with a loincloth. A heavy object should be placed on the abdomen to prevent gas formation and the head should be swathed in a piece of cloth to close the mouth and support the chin (Shadid and Van Koningsveld 1997). Family members or members of the Muslim community generally carry out the ritual washing of the body: men wash men and women wash women. In principle the body is washed three times (or five or seven times) with (scented) water and wrapped in white cloths. The body is then placed in the grave on its right side, with the face in the direction of Mecca (Dijkstra 1995). The body may not be buried in a coffin, nor is it permitted to exhume the remains of the deceased in order to reuse the grave (Dessing 2001). Of course these guidelines refer to what local Muslim leaders stipulate in the context of Muslims in the Netherlands; different or additional Muslim funerary practices may be observed in other parts of the world.

Although Dutch law provides that a body must always be transported in a casket, other burial options have been permitted since 1991 (Van Bommel 2003). Increasing demand – from Muslim refugees and immigrants of Indonesian and Surinamese origin – has led various municipalities to develop Islamic burial facilities, in which the graves are oriented towards Mecca. These are still rarely used by Turkish-Dutch and Moroccan-Dutch families (Willighagen 2010, De Jong 2013), as they are doubtful about municipal assurances regarding grave reuse and visits to the grave (Baba and Gustings 2004). Indeed, an eternal resting place cannot be guaranteed in the public cemeteries in the Netherlands (De Jong 2013). Moreover, many Turkish-Dutch and Moroccan-Dutch families have taken out insurance covering repatriation of a deceased family member to Turkey or Morocco. The funeral directors who are specialised in international burials follow the Islamic requirements around the ritual washing, the transport of the body and the burial in the village of origin and thus ease the material and psychological burden on the bereaved family (De Jong 2013).

The final passage not only implies a very considerable transition for the deceased patient, but also for the relatives, who, moreover, suddenly become dependent on other care providers. This article turns now to examine the diverse experiences of Turkish-Dutch and Moroccan-Dutch families with palliative care professionals and burial care providers, seeking to explain what lies behind these diverse experiences.

Methods

End-of-life care refers to the palliative care provided to a terminally ill patient and their relatives. Burial care refers to the care of the body and the burial of the deceased patient and the care provided to the bereaved. Turkish-Dutch or Moroccan-Dutch persons refer to residents of the Netherlands, who have at least one parent born in Turkey or Morocco.

This article relies on a secondary analysis of data from previous research conducted in 2009–2011 on the communication between Turkish-Dutch and Moroccan-Dutch patients, their relatives and their Dutch palliative care professionals. That research was approved by the medical ethics committee of Zuidwest Holland (no. 07-33) and by the committees of the care organisations involved. The source qualitative research explored the personal experiences of terminally ill individuals, relatives of recently deceased patients as well as professional care providers involved in 33 cases of terminal cancer treatment. The interview topics included the patients' illness, their family situation, the care tasks of family members and professionals, how decisions were made during the palliative phase and beyond, and how the interviewees experienced care, communication and cooperation. The records were typed and qualitatively analysed, leading to codes such as 'personal characteristics', 'family characteristics', 'ideas on good care' (including hygiene, mobility, food, medicines), 'consciousness of being terminally ill', 'rituals in terminal phase', 'care after the patient has died' and so on. For this article, the experiences of families with a Turkish and Moroccan background concerning care and rituals around and after death in the Netherlands were analysed again. Moreover, data were retrieved from participants of focus groups organised to discuss the preliminary findings with more relatives and other Turks and Moroccans. In these focus groups, we discussed topics such as the taboo of disclosing to the patient their terminally ill status; the care in the last days before death; the preferences for dying at home or in hospital; the division of tasks and concerns among the family members and their experiences with care before and after the death of their loved one. These discussions were facilitated by showing participants illustrations like those in Figure 1. Heaton (2008) would classify this secondary analysis as an amplified analysis, as the two sets of self-collected data were combined in order to find an answer to a new research question.

The research population

Information about how the families and the care providers experienced the care provided during and after the final transition is available for 16 of the 33 patients in the original study. Of these patients, 13 have a Moroccan and 3 a Turkish background; 14 patients are first-generation migrants and 2 (nr. 14/7 and 18/9) belong to the second generation. The place of death varied: the exact place of one patient is unknown as he was transported from his house to a hospital in Belgium several times (27/13). One patient died in Turkey (nr. 1/1). All other patients died in the Netherlands: three in a hospital, two in a hospice and nine at home. The main characteristics of the cases can be found in Table 1. Additionally, information was obtained from participants of focus groups organised for Moroccan women in The Hague, in Enschede for Turkish family members and in Noordwijkerhout for Turkish and Moroccan patient representatives, who spoke at length about their experiences with the death and burial of their loved ones.

The transition from care for the ill to care for the deceased: the perspective of the bereaved family

For many families the transition from care for the ill to care for the deceased was a very unfamiliar experience. Especially because dying and death proceeds in a host society and its unknown institutions. They felt they were pioneers in this respect: although they are used to pioneering ever since they left their home country – looking for a job, organising the arrival and housing of their families, surviving economic crises, social conflicts and intergenerational tensions – this pioneering with illness and death felt exhausting and isolating (see also, De Graaff *et al.* 2010b, 2012a, 2012b):

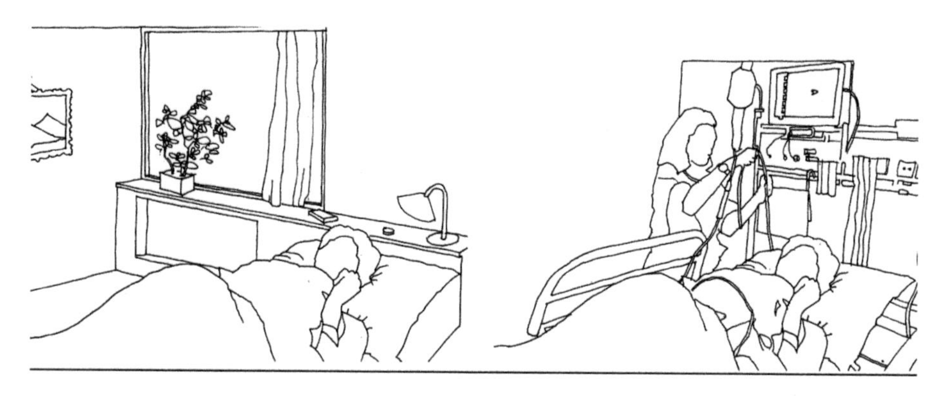

1. At home or in the hospital?
Would you prefer to die in hospital or at home? Would you discuss your preferences with your family and doctor?

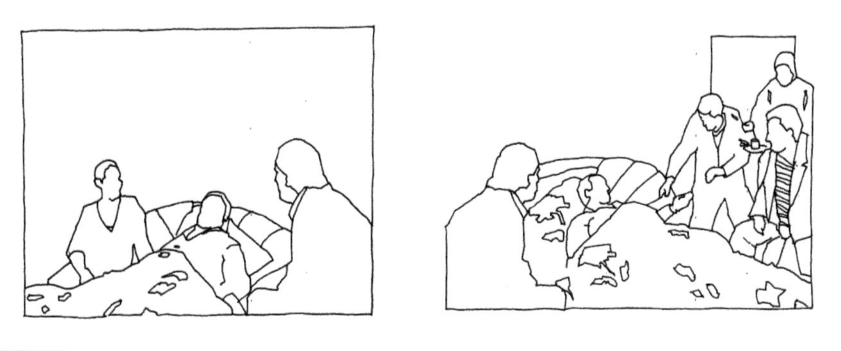

2. Who talks to whom?
Would you prefer the doctor to discuss care decisions with the patient or with the family?

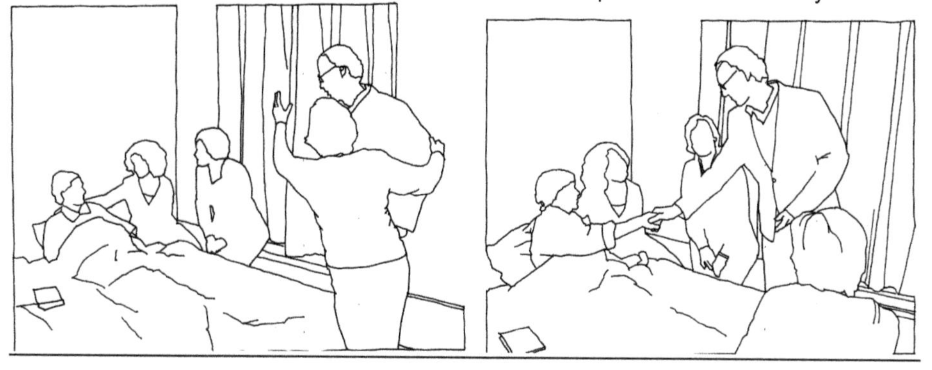

3. The different roles of relatives
How do you combine the interests of the patient and your own interests as a family member? Especially when you are functioning as an interpreter for patient-care provider discussions?

Figure 1. Illustrations used in focus groups.

> The first time I experienced the death of a family member in the Netherlands, it was hard for me. I was isolated. It was the first time we'd experienced something like this. If you'd stayed in Turkey, you'd have had more experience, with grandmothers and uncles. You'd have learned how the process works, but here, we're real pioneers in this respect. (focus group participant, 20100218a)

Table 1. Characteristics of the cases included in secondary analysis.

Case number/art number	Background	Diagnosis	Patient interviewed	Family interviewed	Professional interviewed	Place of death home/in hospital/ in hospice/in home country	Buried in:
1/1	Tu	*Mesothelioma*	no	Daughter	GP	Home country	Turkey
2/2	Mo	Bladder cancer	no	–	GP	At home	Morocco
4/3	Mo	Bronchial cancer	Yes	Wife brother in law	Pain specialist Nurse Home care nurse GP	At home	Morocco
6/4	Mo	Lung cancer	No	Wife	GP	At home	Morocco
8/5	Mo	Stomach cancer	No	Wife	GP Social worker	In hospice	Morocco
9/6	Tu	Brain tumour	No	–	Nurse GP	At home	Turkey
14/7	Mo	Brain tumour	Yes	Mother 2 sisters brother	GP Oncologist pastoral worker	At home	Unknown
15/8	Mo	Lung cancer	No	Daughter	GP	At home	Morocco
18/9	Mo	Stomach cancer	No	Sister	Home care nurse	At home	Morocco
21/10	Mo	Lung cancer	No	Daughter	Social worker Oncologist Pain Specialist Transfer nurse	At home	Morocco
22/11	Tu	Bone cancer	No		GP Home care nurse	In hospital	Turkey
25/12	Mo	Bowel cancer	No	Wife	GP	In hospital	Morocco
27/13	Mo	Stomach cancer	No	Daughter	GP Home care nurse	Unknown	Morocco
28/14	Mo	Brain tumour	No	Daughter	GP	In hospital	Morocco
31/15	Mo	Ovarian cancer	Yes	Daughter	Oncology nurse Home care nurse GP Wound nurse	At home	Morocco
33/16	Mo	Breast cancer	No	Daughter	Nurse Home care nurse	In hospice	Morocco

Note: Tu = Turkish patient; Mo = Moroccan patient.

According to the participants of the focus groups, this lack of experience and uncertainty leads many family members to follow the instructions of their fellow Muslims and community members. This might explain why many bereaved family members and relatives experienced the passage from care for the ill to care for the deceased as a transition from a phase marked by uncertainty and hope – during which heated disputes took place in discussions with care providers about the care that was offered – to a phase of certainty and surrender – in which action was taken swiftly and with full confidence in collaboration with the new care providers. The relations with doctors and nurses were totally different from those with funeral undertakers:

> The weird thing was you could actually see that my father was dying, but still we didn't take it in. He'd been given so much morphine that he was no longer awake. We had a lot of visitors, to the doctors' frustration. But for us, that meant respect. The doctor made me keep visitors at bay, even when they came from far away and I had to tell my mother that she no longer had to wash my father, because the nurses did that. And I had to tell her she wasn't allowed to give him anything to eat, as he could suffocate. He was in a coma for two days and then it was done. I was overwhelmed but relieved, as well. […]. Then I had to wash him, together with my mother. I never thought I would be able to do that. But the strength came from somewhere. My mother and brother washed him, and I fetched all sorts of things for my mother, water, washcloths, sheets. (daughter patient nr 15)

While the relationship between the patient's daughters and the doctors was characterised by negotiations, because they often did not agree with the suggestions made by the care providers, the daughters put up no resistance to the funeral undertakers' orchestration of the proceedings. It felt as if they were partners, since the undertakers demonstrated familiarity with their cultural and religious needs:

> The scans revealed that she had metastases in her head. That affected her behaviour, she didn't recognize us any more, she had a delirium and hallucinations, she was very afraid. The doctor thought it was inhumane to let her suffer like that and wanted to keep her asleep, but the idea that she would never wake up again terrified me. In our religion, it's not allowed for someone to reach the end in that way. They said: it's not euthanasia. I talked it over with my sister and we did not accept their proposal. We stipulated that my mother could be given the sleeping drug at night in order to be able to sleep but that she should be awake during the daytime. I was allowed to sleep in her room in hospital and I read the Qur'an to her the whole time. She died in the afternoon, at one o'clock in the afternoon. It all went very quickly after that, because she was taken straight to the mosque where the washing of the body was performed. I didn't have to arrange that, a contact person did that who came to the hospital on the day she died and who arranged everything. I did help with the washing of the body, together with two other women. We went to Morocco with my aunts and uncles; my mother was insured for that […]. She was buried in Tetouan in the same cemetery as my grandpa. (daughter patient nr 28)

The transition from care for the ill to care for the deceased: the care providers' perspective

The reactions of care providers regarding the transition from caring for the ill to caring for the deceased were varied. For some, this transition demanded a truly pioneering spirit:

> Now, in retrospect, I realize that there were some things that I didn't do right, the way they should have been done. For example, when it came to the ritual washing. On the day that the

> woman died, her husband called his daughter, who then came. I said, what can I do for you? We are accustomed to providing the final care to the residents here, dressing them in clean clothes, just generally cleaning them up. Not in this case. But everything was soiled. The patient had sweated, I said: Would it be all right if we dress her in a clean shirt and clean pyjamas, then she'd look presentable as she lay there, is what we thought. Well, we were allowed to, and so we did. But they left before that. He walked away. I said, would you like to wait, you can get some coffee and see her again when she's been tidied up. No, that wasn't necessary. Now, I understand that that's part of the culture. When the undertaker came to fetch her, her husband did not accompany her inside. The funeral undertaker called the morgue in Utrecht where she was ritually washed by two women. She was then taken to the refrigeration cell at Schiphol, where her husband and daughter were to catch a plane. They were insured for that. (hospice nurse of patient nr 33)

Some of the more experienced care providers remarked that they felt pushed aside during this phase; they were not involved in any of the care-related choices; the family simply went its own way:

> The patient was admitted urgently to the hospice in the afternoon, was restless and was given a morphine pump. His wife spent the night. The notes read: 'the family is Muslim and observes Ramadan', in other words, as soon as it's light, don't offer anything. They also say that he will be ritually washed in a mosque and then immediately flown to Morocco, where he will be buried. 'Do nothing, no candle', it says. Okay. They're Sunni Muslims, they don't burn candles, if they're Shiites, they do burn candles. Alewites have a whole bunch of different traditions, and Moroccan Muslims are totally different from Turkish Muslims. So what I know is, they can say prayers after death has occurred, it's a whole ritual. And also that burial is very essential in order to depart well to the hereafter. But I felt pushed aside by their fear that we'd do something wrong. Because I came the day after, when he was already dead. I went upstairs, to the son. And I expressed my condolences. And then I said, look, I understand that you're arranging everything yourself? Yes. I said: when someone leaves the hospice we always hold a very small gathering together, in any case to show respect. Well that was out of the question. I said it is important for us as well, to have that chance to say goodbye. Because we took care of your father. No, it was out of the question. I thought it was a real shame. There was no communication, they walked away, just like that. It all went very fast. (social worker of patient nr 8)

Other care providers had developed such a bond with the family that they were able to interpret the needs of the various family members, including their wishes regarding the final transition. They were neither surprised, nor disappointed:

> I feel you should give people the opportunity to make their choices without losing face. I explained a great deal to this family, I visited them at home but I also arranged for first the men and then later the women to come in to the office so that they could absorb the necessary information. He developed decubitus, the family thought that that was part of the process of dying, but we could remedy that. He didn't want any painkillers. I told him that it pained me to see him suffering like that, but he wanted to stay in control. I explained to them what would happen after he died . . . , I showed them the forms, what needed to be filled in, the questions these contained, to avoid overwhelming them when the time came. I also asked them whether the funeral had been arranged, who they needed to call. Yes, all that was sorted. So, when on the Saturday before he died someone from the mosque came in when I was there, I asked: Would you like me to leave you alone for a moment? But no, I was thoroughly introduced and was allowed to remain in the room. It was a super experience, even now I still get tears in my eyes from it. (GP of patient nr 2)

These examples clearly indicate that differences in reactions are not only typified by a difference in awareness of Islamic funeral rituals, but above all by the extent to which professionals and clients are sharing their views on problems and solutions and their management of care options. Even afterwards care providers seldom review their experiences with these relatives. Many care providers make a point of contacting the bereaved family after the death to offer aftercare, if desired. Yet because the Turkish and Moroccan families usually travel to the country of origin at that time, the aftercare visit is often skipped, which means that any later sharing of knowledge and insights also fails to take place:

> I did not contact them after his death anymore. I usually visit families of deceased patients to evaluate and give aftercare, but they stayed away [in Morocco] for a long time. (home care nurse of patient 9/6).

Common experiences indicating structural features of end-of-life care and care of the deceased

First it is worth noting that all respondents emphasised that, as a rule, dying and burial are different for Turks and Moroccans compared with Dutch people's practices. These differences are not defined in terms of individual or family features, but in terms of culture and religion:

> In our culture and according to our faith, when someone dies, we have to undress him right away and wrap him in sheets and bury him as quickly as possible, we don't want him lying around on ice for a few days. (daughter of patient 1f)

A second point is that both care providers and families were afraid of performing the 'wrong actions'. Family members visited the hospital to cheer up the patient, but also to check on the care being provided. They noted that the hospice smelled of candles, as if it were a church and that the care there was organised according to traditions over which they had no control. This lack of control, however, is rarely discussed:

> I saw that there were a great many spiritual caregivers, that sets you thinking. I asked them: What do you do when you give care to a Muslim patient, do you have an Imam who helps out? Because we don't know anything about medicines, we're afraid .... To us it's important to have a clear mind at the time of death, to be able to bear witness. We are afraid that the morphine will cause a person to die faster, and we don't want that. (focus group participant 20100218c)

Some care providers at times also feared that the family would not provide adequate care. Other care providers acknowledged that their care was directed at physical solutions and less at sharing psychosocial insights:

> What the wife's role was? Well, she cared for him, only I wasn't able to communicate with her very well, her Dutch was extremely poor. I wondered whether she might not be mentally impaired. But when I asked the director of the homecare agency about that, she said, not at all, that lady makes a perfectly adequate impression in Turkish. So it's just the language, the language barrier makes it hard to judge. (GP of patient nr 1)

> If you really want choices to be made together, it's not just about the solutions, but also about the analysis of causes and emotions. Do we perceive the same causes, and what fears are involved? She, herself, is not afraid, her trust in Allah gives her comfort. But her son has hyperventilation symptoms, and fears for his heart. Death is soon approaching and for

> him, and many others, that's a frightening situation. But I didn't discuss that, there are so many missing pieces and things that in a different situation [in the case of Dutch families] you naturally would discuss. (GP of patient nr 31)

A final noteworthy point is that the care provided prior to death would appear to be far more of a burden on the family than the care after death. This is probably due to the fact that palliative care professionals must communicate in triads, as they serve both the patient and the members of the patient's family. On the contrary, undertakers communicate only with the family. Even then making decisions can be a fraught task, as the bereaved family members may have different wishes. Yet the tasks associated with a funeral are well-defined and the expenses are in many cases partially covered by the insurance, while the process of dying is much less straightforward. However, this lack of straightforwardness seems to stem from the different relationship with the service providers: the Turkish and Moroccan families trusted the funeral undertakers because they were cooperating with migrant associations and the Turkish and Moroccan insurance companies. The undertakers ensured their services were fully aligned with Islamic rules and they employed bilingual staff to guarantee care that was made-to-measure. In contrast, the relationship with doctors and nurses was characterised by a continuous process of negotiation about the care provided, in which the Dutch perspective on good care took the lead. No longer having to negotiate with Dutch care providers had the additional effect of restoring the old, familiar roles in families: they regained a sense of control, in a situation where language fluency was no longer a main concern and the use of traditional rituals affirmed the familiar roles:

> Well, I decided my mother would go into the hospice. My father was apprehensive about the costs, but I said: I'll pay. Then my father was worried because she was no longer receiving certain drugs. I contacted the hospital, the GP and the nursing staff about that, and explained to my father that the drugs no longer worked. That was obviously a problem with the language barrier. I was present when, very peacefully, she died. Um, and … then my father decided – those were decisions taken by my father – he wanted to wash her and prepare her, but we had to wait until death was pronounced by the doctor. My father then said, well then we'll leave, because he didn't want to see her again, like at that point it was finished for him, okay, she's gone and now we're going home. She was then transported to Utrecht, where she was formally washed. A whole lot of Moroccan families, probably Turkish ones, as well, are insured at the same insurance company for when somebody dies. Most Moroccans are insured at the Banque Populaire. We called this bank, and they contacted the undertaker in Utrecht. It meant my father had a lot less to worry about. The undertaker arranged that everything could be done in Utrecht, as was proper. He also took care of the tickets, so that we could both fly to Morocco. My aunt took care of everything in Morocco. (Daughter of a Moroccan patient nr 33)

Considerations and conclusions

The experiences of Turkish and Moroccan families and their Dutch care providers in end-of-life care and beyond clearly indicate cultural differences. Moreover, they demonstrate structural features in power relations and logics of care. The observation that more tensions arise during the palliative care phase than during the phase of burial care can lead to new insights into the logics of care of medically oriented palliative care professionals and service-oriented undertakers. After all, the habitus of medical professionals is different in the sense that they (traditionally) claim to know what is best for patients and relatives, while undertakers are mainly guided by the commercial imperative to satisfy their clients' wishes. Recent research on the different logics of care (Verhagen 2005, Knijn and Verhagen 2007) has yielded useful concepts to interpret the different patterns of tensions between, on the one hand, care professionals and their patients and the families of those patients, and on the other hand, undertakers and their clients.

According to Knijn and Verhagen (2007), the relationships between care professionals and their patients are embedded in a care logic of professionals who are politically authorised to dispense advice and treatment, because of their knowledge of universal medical relief standards and expertise in health skills. This expertise results in a relationship with their patients that is characterised by power and intimacy at a distance. The relationship between the family members of a patient and the patient themselves is embedded in a community or family-based care logic, characterised by a personal bond and informal solidarity. Finally, the care logic around the burial has both characteristics of professionality (treating the deceased person with care and paying attention to the needs of relatives) and characteristics of 'customer care' expressed by a relationship between an entrepreneur and his/her consumers, which follows the logic of the marketplace. These different care logics lead to predictable annoyances. In conformity with Becker, in his book *Tricks of the Trade* (Becker 1998), and Verhagen (2005), the fact that fewer tensions arise between undertakers and their immigrant clients is unsurprising, as both adhere to care logics functioning in the private as opposed to the public sphere. By contrast, the professionals operate in the politically authorised public sector. In other words, undertakers and other commercial care organisations can be more flexible in adapting their services to new clients and in using suitable subcontractors, in order to satisfy the range of different needs. Given that operating in the private/market sector is also accompanied by a host of disadvantages, this analysis of care logics is not intended as an argument in favour of commercial palliative care. Yet, health care organisations might well benefit from embracing a few of the policies that have already been adopted by the professional services sector, and more specifically, by funeral homes; a functional collaboration with mosque organisations, an unhesitating adaptation to Islamic rules and the availability of bilingual employees to provide this specific type of care have all proven to result in better communication and greater confidence in the care provided among Turkish–Dutch and Moroccan–Dutch bereaved families, while at the same time contributing to improved family relations, which is obviously another goal of palliative care.

Additionally, it is important not to label cultural differences as Turkish/Moroccan traditions versus Dutch traditions, or Christian versus Muslim rituals, but to explore the state of mind of the individual (and their family) in question. Not only do uniform Islamic guidelines go hand in hand with variation in the practices of Muslims (Harmsen 2007, Venhorst 2012), there is also no uniformity among the Dutch palliative patients and their burials. Hence, palliative care professionals are seeking to align their universal standards with the specific needs of each patient and family (WHO 2015), as this discretionary authority to interpret such standards is the core of their professionalism (Freidson 2001).

Further, care providers can break the taboo of discussing death and burial in some families, not by imposing their own solutions, but by exchanging information. They can inform the family about the many different alternatives espoused in the traditionally compartmentalised society of the Netherlands (Phalet and Ter Wal 2004) and ask about the family's customs (De Graaff *et al.* 2012b) and the desired care after death. Not in the role of fellow decision-maker or orchestrator, but simply as an interested bystander and fellow citizen (Mistiaen *et al.* 2011). Sharing knowledge and views on end-of-life care management after the funeral not only offers family members support in how to cope without their loved one, it also affords palliative care professionals with an opportunity to learn how to honour the personal wishes of culturally diverse patients and their families within the boundaries of universal medical relief standards for 'end-of-life care and beyond'.

As this study is solely based on the experiences of the bereaved relatives and palliative care providers, the findings can only be considered as indicative. Further research into the experiences of service providers in the funeral industry is needed to obtain a clearer idea of the similarities and differences in the care provided by the two groups and the (possible) learning points that can be gleaned from the approaches of both. It is also important to realise that the practice of end-of-life care is changing swiftly, as the migrant pioneers in this area are blazing the trail for an upcoming cohort.

References

Baba, M. and Gustings, F., 2004. *Wat is uw laatste wens? Een verkennend onderzoek naar de behoefte, wensen en ideeën m.b.t. islamitische begraaf-gelegenheden onder Amsterdamse moslims [What about your last wishes? Explorative research on the needs, wishes and ideas about Muslim burial opportunities of Muslims in Amsterdam]*. Amsterdam: Mexit.

Becker, H.S., 1998. *Tricks of the trade: how to think about your research, while you're doing it.* Chicago: University of Chicago Press.

Bot, M., 1998. *Een Laatste Groet: Uitvaart- en rouwrituelen in multicultureel Nederland [A last greeting: rituals for funerals and mourning in the multicultural Netherlands]*. Rotterdam: M. Bot.

Buiting, H., *et al.*, 2008. A comparison of physicians' end-of-life decision making for non-Western migrants and Dutch natives in the Netherlands. *European journal of public health*, 186, 681–687.

Bulow, H. *et al.*, 2008. The world's major religions' point of view on the end-of-life decisions in the intensive care unit. *Intensive care medicine, 34*, 423–430.

Daar, S.A. and Khitamy B.A., 2001. Bioethics for clinicians. 21. Islamic bioethics. *Canadian medical association journal*, 33, 60–63.

Denktaş, S., 2011. *Health and health care use of elderly immigrants in the Netherlands.* Rotterdam: Erasmus University.

De Graaff, F.M. and Francke, A.L., 2003. Home care for terminally ill Turks and Moroccans and their families in the Netherlands: Carers' experiences and factors influencing ease of access and use of services. *International journal of nursing studies*, 40, 797–805.

De Graaff, F.M., *et al.*, 2010a. *Communicatie en besluitvorming in de palliatieve zorg voor Turkse en Marokkaanse patiënten met kanker* [Communication and decision making in palliative care for Turkish and Moroccan cancer patients]. Amsterdam: University of Amsterdam.

De Graaff, F.M., *et al.*, 2010b. 'Palliative care': a contradiction in terms? A qualitative study of cancer patients with a Turkish or Moroccan background, their relatives and care providers. *BMC Palliative Care*, 9 (19), [online].

De Graaff, F.M., *et al.*, 2012a. Talking in Triads: communication with Turkish and Moroccan immigrants in the palliative phase of cancer. *Journal of clinical nursing*, 21, 3143–3152.
De Graaff, F.M., *et al.*, 2012b. Understanding and improving communication and decision making in palliative care for Turkish and Moroccan immigrants: a multiperspective study. *Ethnicity and health*, 17 (4), 363–384.
De Graeff, A., *et al.*, 2012. Palliatieve zorg voor mensen met een niet-westerse achtergrond: een handreiking met adviezen, [Palliative care for non-western immigrants: helpful advices]. *NTPZ. Nederlands-Vlaams Tijdschrift voor Palliatieve Zorg*, 12 (2), 4–20.
De Jong, F., 2013. *Islamitisch begraven in West-Brabant: Een onderzoek naar de wensen en mogelijkheden* [Muslim burials in West-Brabant:Tilburg: A study about the wishes and opportunity's]. Tilburg: Kennisklik, Tilburg University.
Dessing, N.M., 2001. *Rituals of birth, circumcision, marriage and death among Muslims in the Netherlands.* Leuven: Uitgeverij Peeters.
Dijkstra, C., 1995. *'s Lands wijs, 's lands laatste. eer, over culturen and overlijden.* [When in Rome, bury as the Romans do, about culture and death]. Rijswijk: Centraal Orgaan opvang asielzoekers.
Freidson, E., 2001. *Professionalism, the third logic. On the practice of knowledge.* Chicago: The University of Chicago Press.
Harmsen, P., 2007. *Handleiding Islamitisch Begraven* [Manual Muslim burying]. Den Haag: Sdu Uitgevers.
Heaton, J., 2008. Secondary analysis of qualitative data: an overview. *Historical social research*, 33 (3), 33–45.
Knijn, T. and Verhagen, S., 2007. Contested professionalism. Payments for care and the quality of home care. *Administration and society*, 39 (4), 451–475.
Mistiaen, P., *et al.*, 2011. *Handreiking Palliatieve zorg aan mensen met een niet-westerse achtergrond en Achtergronddocument* [Aid Palliative care for non-Western patients with Background document]. Utrecht: NIVEL.
Onwuteaka-Philipsen, B., *et al.*, 2007. Physician discussions with terminally ill patients: a cross-national comparison. *Palliative medicine*, 21, 497–502.
Phalet, K. and Ter Wal, J., 2004. *Moslim in Nederland* [Muslim in the Netherlands]. The Hague: Sociaal Cultureel Planbureau.
Shadid, W.A.R. and Van Koningsveld, P.S., 1997. *Moslims in Nederland* [Muslims in the Netherlands]. Houten: Bohn Stafleu Van Loghum.
Van Bommel, A., 2003. Het Islamitisch sterven. *In*: M. Pijnenburg, eds. *Multicultureel sterven in het ziekenhuis.* Budel: Damon, 82–93.
Venhorst, C., 2012. Islamitic death rituals in a small town context in the Netherlands: explorations of a common praxis for professionals. *OMEGA*, 651, 1–10.
Venhorst, C. 2013 *Muslims rutualising death in the Netherlands. Death rites in an small town context.* Nijmegen: Radboud University Nijmegen.
Verhagen, S., 2005. *Zorglogica's uit balans. Het onbehagen in de thuiszorg nader verklaard* [Struggling logics of care. Home care and its discontents explained]. Utrecht: Uitgeverij de Graaff.
VPTZ, 2008. *Gaat u het gesprek aan? Goede zorg voor stervenden van allochtone afkomst. Eindrapport.* [Does this conversation concern you? Good quality care for dying immigrants. Final Report] Bunnik: VPTZ.
Vroon-Najem, V., 2014. *Sisters in Islam, women's conversion and the politics of belonging, a Dutch case study.* Amsterdam: University of Amsterdam.
Willighagen, G.H. 2010. *Islamitisch begraven* [Burying the Muslim way]. Zwolle: Onderzoek & Statistiek, Afdeling Informatie.
World Health Organisation (WHO). 2010. *Definition of palliative care.* Available from: http://www.who.int/cancer/palliative/definition/en/ [Accessed 7 January 2016].
World Health Organisation (WHO). 2015. Palliative Care Factsheet, nr 402. Available from: http://www.who.int/mediacentre/factsheets/fs402/en/ [Accessed 7 January 2016].
Yerden, I., 2000. *Zorgen over zorg. Traditie, verwantschapsrelaties, migratie en verzorging van Turkse ouderen in Nederland* [Concerns about care. Tradition, relationships, migration and the care of elderly Turkish migrants in the Netherlands]. Amsterdam: Het Spinhuis; 2000.

Yerden, I., 2004. Blijf je in de buurt? Zorg bij zorgafhankelijke Turkse ouderen 2004. [Can you stay? Care for care-dependant elderly Turkish immigrants]. *Cultuur Migratie Gezondheid*, 1, 28–37.

Yerden, I. and van Koutrike, H., 2007. *Voor je familie zorgen? Dat is gewoon zo. Mantelzorg bij allochtonen. Mantelzorg bij Antillianen, Surinamers, Marokkanen en Turken in Nederland* [Caring for your family? That's just something you do. Care for ethnic minority patients by family and friends. Care for immigrants in the Netherlands from the Antilles, Surinam, Morocco and Turkey]. Purmerend: Primo.

Between Civil Society and the State: Bureaucratic Competence and Cultural Mediation among Muslim Undertakers in Berlin

Osman Balkan

Department of Political Science, University of Pennsylvania, Philadelphia, PA, USA

ABSTRACT

This article explores the intercultural negotiations around the death and burial of Muslims in Germany. In particular, it examines the mediating role that Muslim undertakers play between immigrant families and the German state. Drawing on an ethnographic study of Turkish funeral homes and the Islamic funeral industry in Berlin, it argues that undertakers' ability to navigate the regulatory structures of the German bureaucracy and the cultural expectations of their customers is a defining feature of their occupational identity and a principal source of their professional authority. As intermediaries between civil society and the state, undertakers guide families through the cultural, religious, political, and legal landscapes that structure the transitions from life to death. In burying the dead and tending to the living, they must reconcile competing sets of administrative and cultural norms surrounding death and interment. In doing so, the Muslim undertakers of Berlin preside not only over end-of-life decisions and their theological implications, but also over pedagogical moments in processes of political and cultural integration in contemporary Germany.

Undertakers occupy a unique niche in the world of professions. Although they operate within the legal parameters of the market, their proximity to death and the disquieting idea that their livelihood is based on the grief and suffering of others can be a source of stigmatisation (Thompson 1991, Cahill 1995). Conversely, their ability to help guide families through difficult and often painful situations can earn undertakers a great deal of respect and admiration from the communities they serve (Laderman 2003, Smith 2010). While their primary task is the disposal of the dead, a central component of undertakers' professional responsibility involves attending to the living. In this capacity, they assume different roles and perform a wide range of activities from bereavement support and religious counselling to legal arbitration and conflict resolution (Lynch 1997). The work of undertaking takes on political salience in multicultural settings where different ethnic and religious groups have divergent views on death and dying, end-of-life care, and the proper treatment of corpses. In situations where there is some uncertainty about funerary traditions and burial laws, undertakers often play a critical part in navigating the political, legal, religious, and cultural landscapes that structure transitions from life to death.

This article explores the intercultural negotiations around the death and burial of Muslims in Germany. In particular, it examines the mediating role that Muslim undertakers play between immigrant families and the German state. The rites and rituals associated with death are remarkably varied across cultures, as are the laws and institutions that regulate the governance of dead bodies. When a death occurs in migratory situations, families are often compelled to negotiate alternative systems of burial and memorialisation (Oliver 2004). The laws of the dead can be at odds with the cultural traditions and religious beliefs of immigrant groups, leading to conflicts around the handling of corpses (Renteln 2001, Carpenter, B., et al., 2015). Moreover, death rituals themselves might undergo change in migratory contexts when immigrants encounter different ways of performing funerary rites and adapt their own practices in response to institutional constraints in the host society (Venhorst 2012).

Immigrant families are faced with a number of other choices, including the method of disposal (burial, cremation, or other alternatives) and the site of the funeral (in the country of origin, settlement, or both). Such decisions are influenced by various factors such as costs, territorial attachments, the availability of appropriate burial grounds, family ties, and feelings of social exclusion (Attias-Donfut *et al.* 2005, Balkan 2015a). In recent years, a growing number of private companies, governmental agencies, and funeral funds have been established to subsidize, facilitate, and in some cases, encourage the transportation of corpses and human remains both within and across international borders (Félix 2011, Marjavaara 2012, Jassal 2014, Balkan 2015b). The increased mobility of the dead has led to a proliferation of new strategies for mourning and memorialisation that may or may not draw on existing repertoires of grief (Prendergast *et al.* 2006). By augmenting the period between death and burial and extending it over geographical space, transnational funerals have impacted both the ritualization of death and the meanings attributed to the place of burial. While this article is less concerned with broader transformations in mortuary rituals in the wake of heightened mobility (see Gardner 2002, Mazzucato *et al.* 2006, Zirh 2012), it is important to point out that undertakers play an important role in this process, particularly as intermediaries between grieving families and the state.

When Muslim migrants with different social and sectarian backgrounds disagree about what constitutes a proper Islamic funeral, undertakers are able to intervene to bridge the gap between what is permissible and what is possible under Islamic law. Previous research on the funerary rituals of Muslim minorities in Germany has shown how Muslim undertakers mediate religious disputes by developing practical solutions to seemingly unsolvable theological problems (Jonker 1996). Although they usually lack the requisite training to make authoritative pronouncements on religious questions, their expertise in matters related to the dead provides Muslim undertakers in the diaspora with a certain amount of religious credibility. As such, they are able to offer religious guidance to Muslims in Germany and to mediate disagreements that arise within the community.

While knowledge of different religious and sectarian traditions is critical to the provision of appropriate funerary services, this article contends that a different type of mediation is also central to the work of undertaking in a migratory context. As private actors who negotiate issues of citizenship and sovereignty in relation to the dead, Muslim undertakers serve as political and cultural mediators between immigrant families and the German state. In the following analysis I argue that their ability to navigate the regulatory structures of the German bureaucracy and the cultural expectations of their customers is a defining feature of their occupational identity and a principal source of their professional authority. In burying the dead and tending to the living, Muslim undertakers must reconcile competing sets of administrative and cultural norms surrounding death and interment. In doing so, they preside not only over end-of-life decisions and their theological implications, but also over pedagogical moments of political and cultural integration in contemporary Germany.

The rest of this article proceeds as follows. I first provide an overview of the German funeral industry and the legal structures governing burial in Germany. I then move to a description of my field site and methods. This article draws on ethnographic and qualitative research conducted among Turkish undertakers who provide Islamic funerary services in Berlin. I present and analyse narratives drawn from my interviews to show how the undertakers offer political and cultural mediation between civil society and the state. I focus on two dimensions of their mediation that relates to their knowledge of the German bureaucracy and their efforts to combat stereotypes about Muslims in Germany. I conclude with a discussion about the role of undertakers in intercultural negotiations around death and dying.

Becoming an undertaker

The German death-care business is estimated to be a 15 billion euro industry (Akyel 2013). Although there are some larger companies that have subsidiaries throughout the country such as the well-known *Ahorn-Grieneisen*, it is a sector that is mostly dominated by small, family-owned firms with a long company history (*Ibid*). According to the website of the Federal Association of German Undertakers (Bundesverband Deutscher Bestatter e.V., 2015), there are approximately 4000 funeral homes in Germany that oversee the burial and cremation of 860,000 people annually (Bundesverband Deutscher Bestatter e.V., 2015). While it is difficult to estimate the total number of Islamic funeral homes in Germany, they currently represent a small share of the overall market. In the city of Berlin, which is the focus of this study, there are around 300 funeral homes, 6 of which are explicitly oriented towards the city's Muslim communities.

Although there are no educational requirements to become an undertaker in Germany, individuals wishing to enter the funeral industry can receive hands-on training through Germany's vocational education system (*Berufsschule*). In 2003, undertaking was added to the list of 350 officially recognized professions that vocational students can apprentice in. The creation of the first Federal Training Center for Undertaking in the Bavarian city of Münnerstadt in 2005 has enabled death workers to receive further training in mortuary sciences. The Centre offers one- to three-year educational programmes that cover topics such as grief psychology, business administration, and coffin construction. While the goal of the Centre is to help institute standardized training in the funeral industry, in practice, little more than a business license is necessary to open a private funeral company in Germany.

Consumers can freely purchase funerary services through the commercial marketplace, but the handling and disposal of corpses is strictly regulated by the German state. Up until the 19th century the organization of funerals in Germany as in other parts of Europe was primarily the domain of the Church (Kselman 1993, Walter 2005). With the growth of towns and cities and concomitant problems of sanitation and disease, the state took a more active role to contain the public health risks posed by dead bodies by creating new institutions to oversee the burial of the dead. Cemeteries were placed under the control of municipal authorities and comprehensive burial laws were established at the state level, dictating everything from the depth and width of a grave to the number of years that a burial plot could be leased before it is recycled and reused.

As Schulz (2013) has observed, modern German sepulchral culture is highly regulated and regimented. The decentralized nature of German burial law means that there are important discrepancies in terms of what burial practices are legally permissible across Germany's sixteen states. Schulz (2013) points out that six states prohibit the use of open caskets at funeral ceremonies; two states prohibit burial at sea; and only one state allows the establishment and ownership of cemeteries by non-municipal or non-religious entities. All states require that corpses and cremains are buried in public or officially licensed cemeteries, which means that practically speaking, it is illegal to keep ashes in one's home. Moreover, only five states allow for burial without a coffin (shroud burial) in line with Islamic and Jewish traditions.

Having an authoritative command of burial law helps constitute the undertaker as an expert. Planning a funeral can be a bewildering and emotionally draining process and one of the ways that undertakers alleviate some of the uncertainty surrounding funerary proceedings is by advising families on the rules governing burial. Their ability to provide accurate answers to questions about interment and to explain the procedural steps leading up to the funeral endows them with professional credibility. This is all the more evident in situations where people have different assumptions and expectations about what a funeral entails.

There are a number of important discrepancies between Islamic funerary traditions and the laws of the dead in Germany that can create confusion and cause problems for families and undertakers alike. According to Islamic tradition, a corpse should be washed, shrouded, and buried as soon as possible in a grave facing the Qibla in Mecca. The dead are to be buried amongst other Muslims rather than in mixed-confessional parcels and the grave should be left undisturbed in perpetuity. In practice, Muslims in Germany face a number of legal and practical impediments in the mortuary realm, including mandatory waiting periods of 48 hours between death and burial, time limits ranging from 5 to 20 years on the leasing of grave plots, the obligatory use of a coffin for burial in 11 out of 16 federal states, and the limited availability of Islamic cemeteries or sections of cemeteries that are reserved exclusively for Muslims. Of the approximately 32,000 cemeteries in Germany, only 250 – less than 1 Per cent – have parcels that are reserved for Muslim graves (Initiative Kabir 2015).

Since the 'success' of a funeral ceremony is judged by the degree to which it conforms to customer expectations, undertakers must balance the interests and desires of bereaved families with the laws of the state. If they promise more than they can realistically deliver, they run the risk of harming their reputation and losing future business. Conversely, if they appear too rigid and inflexible, they may be viewed as insensitive and criticized for acting against the interests of families.

Method

This discussion is situated within a larger research project on Islamic deathways in Germany that explores variations in how death is managed and memorialized by ethno-religious minorities in a migratory context. My primary research site was Berlin, a city with roughly 3.5 million inhabitants, 15.3 per cent of which are foreign nationals and 30 per cent of whom have a 'migration background' according to figures from the most recent census (*Statistik Berlin Brandenburg* 2014).[1] Berlin is not only one of the most ethnically diverse cities in Germany, it is also a place where migrants are more actively involved in local politics relative to other German cities, particularly on issues related to integration and citizenship rights (Koopmans 2004). As such, it is an opportune site to investigate the link between burial practices and political integration.

Over the course of a four-month period in 2013–2014, I conducted 40 interviews with Turkish and Kurdish families, Muslim undertakers, cemetery personnel, religious leaders, government officials, health-care professionals, and representatives of funeral assistance funds. In selecting my interview partners, I was guided by contacts at mosques and cultural centres that cater to a wide range of Berlin's immigrant community, including members of Sunni, Shi'a, and Alevi faiths, and individuals with Turkish and Kurdish backgrounds. My fieldwork included extensive participant observation of the Islamic funeral industry in Berlin, a sector that is largely dominated by Turkish firms. Out of the six officially registered companies that offer Islamic funerary services, five are owned and operated by Turkish Muslims. I spent a significant amount of time shadowing these undertakers and observed every aspect of their work. This included picking up corpses from morgues and hospitals, preparing them for burial, and transporting them to the cemetery or to the airport for international shipment, visiting different municipal offices and foreign consulates to acquire and file the requisite paperwork for burial or repatriation, attending funerals and funeral services at mosques, and observing interactions with customers, families, religious leaders, and other service providers such as cemetery workers. In total, I shadowed seven undertakers from five of Berlin's six Islamic funeral homes. I was unable to meet with anyone from the sixth funeral home, in spite of repeated attempts to make contact.

Lengthy semi-structured interviews were conducted with 6 of the undertakers, all of whom were Turkish or Turkish-German men between the ages of 35 and 60. Two of them were born and raised in Berlin, two emigrated with their families as teenagers, and two came to Germany as adults. None of them had any previous experience or had a family connection to the funeral industry before becoming an undertaker in Germany. Collectively, they had worked in the funeral business for between 5 and 30 years with an average length of 18 years. While the number of women in the funeral industry has increased in recent years, the division of labour tends to be highly gendered and women often hold administrative roles, whereas men are tasked with handling, transporting, and preparing the corpse for burial (see Parsons 1999, Bremborg 2006, for exceptions see Pringle and Alley 1995, Doughty 2014). Of the five Islamic funeral homes that I studied, only one had a female employee and her primary role in the company was as an accountant. Although my own positionality as a Turkish-speaking male may have helped earn the trust of my informants and positively impacted their willingness to participate in this study, several of the undertakers mentioned that they had been previously interviewed by other researchers, both men and women and Germans and Turks alike. Most of them were enthusiastic participants and were eager to share their stories and experiences.

Many insights were gleaned from unprompted comments and interactions in the field, but the more formal setting of the interview allowed me to collect data in a more systematic manner. All of the interviews took place at the undertakers' place of work and lasted between one to four hours. They were often followed up with clarifying questions at a later date. The interviews were conducted in Turkish and translated, transcribed, and thematically coded by the author. The names of the undertakers have been changed to protect their identities. A broad range of topics were covered, including their own personal histories and entry into the funeral industry, the laws of the dead and the feasibility of Islamic burial in Germany, differences between Christian and Islamic funerary traditions and funeral companies, and the attitudes and expectations of their customers. When the theme of cultural and political mediation emerged as a dominant and recurring trope in the transcripts, a series of sub-themes were identified, including bureaucratic competence, socio-political integration, and combatting stereotypes about Muslims. The discussion that follows will largely focus on these three issues.

The timing of this research project proved to be auspicious. In May 2014, two weeks before I returned to the field, German police raided the offices of an Islamic funeral home and eighteen other properties across Berlin in a coordinated effort to crack down on a network of human traffickers. Although the criminal investigation is still ongoing, two Muslim undertakers as well as a city official were taken into police custody and questioned about their alleged involvement in the sale of passports of the dead to human traffickers in Syria and Palestine (see New York Times 2014). This scandal offered an opportune moment for members of the Islamic funeral industry to reflect on the behaviour of their colleagues and competitors and to offer a defence of their own professional integrity.

In the next section, I draw on these narratives to analyse the relationship between bureaucratic competence, cultural know-how, and professional authority. As stated above, an important dimension of the undertakers' authority is premised on their familiarity with the laws surrounding death and their capacity to anticipate and manage customer expectations. In their dealings with bereaved immigrant families, a lack of knowledge about the German bureaucracy was taken as a sign of poor integration. The undertakers' ability to mediate between civil society and the state enabled them to assert jurisdictional boundaries and to claim professional status.

Bureaucratic competence

'It's not like Turkey here', says Mert, lighting a cigarette and inhaling deeply. We are sitting in his office, located on a busy commercial strip in the neighbourhood of Neukölln. The funeral home is situated on a block that is lined with restaurants, cafes, hookah bars, bakeries, supermarkets, and travel agencies. Most of the signage is in Arabic or Turkish, reflecting the demographic composition of the neighbourhood. Around a quarter of Neukölln's residents are immigrants, hailing from many different countries in the Middle East. Walking around the neighbourhood, some might quibble with Mert's assessment given the conspicuous signs of ethnic identity, but

Mert is not speaking about the cultural topography of his block. His comments pertain to the German way of death.

In confronting death in the diaspora, immigrants are compelled to navigate different bureaucratic structures, burial practices, and rituals of memorialization that are incongruous and potentially antithetical to the rites and traditions in their country of origin. In such situations, undertakers play an important pedagogical role. They must instruct their customers not only about the laws of the dead but by extension, the legal-rational order of the host society. 'Our people have been here for fifty years and they still think they can ship a body on a Saturday or a Sunday', Mert continues. 'They think this is like some village in Turkey. But there are a lot of formalities here. We haven't been able to teach this to our people'.

The notion of the Anatolian village was a recurrent theme in my interviews. Turkey, or some idea of Turkey, informed much of the undertakers' judgements and observations about the German system. Some, including Mert, had spent time between both countries and drew on their own personal experiences when making assessments about the differences between the two countries. Bora, another undertaker whom I interviewed, has been in the funeral business for eighteen years. He emigrated to Berlin with his parents from a village near Ankara at the age of 11. While his childhood was spent in Turkey, most of his formative years were in Germany. Nonetheless, the Anatolian village serves as a major point of reference during our conversation.

'Our people have been living here for fifty years', he tells me.

> They could live here for another hundred and fifty years and they still wouldn't understand the system! They don't understand the German system, nor do they want to understand it. Whatever pre-existing mentality they brought with them from Anatolia, that village mentality, it's still there.

Although we are discussing burial practices and the German bureaucracy, Bora and Mert emphasize something much broader. They insist that their customers have unrealistic expectations that indicate poor integration into German society. Both men refer to the half-century of Turkish migration to Germany and speak to what they see as a lack of acculturation to German norms. For Mert, a customer who assumes that a corpse can be repatriated on a Saturday is clearly unfamiliar with the working hours of state offices in Germany, which are closed on the weekends. Such assumptions are seen not merely as misunderstandings, but as evidence that some migrants stubbornly refuse to adapt to the German order. In a similar vein, Bora posits that Turkish immigrants have anachronistic ideas about the way things work in Germany owing to their 'village mentality'. Bora is critical of his customers for what he sees as their unwillingness to acknowledge or accept the structural constraints of the 'German system'. Their mentalities are seen as durable and portable dispositions, not unlike a habitus, that frames the ways in which they approach and navigate German space.

Even undertakers who have little personal experience in the ways of rural life in Turkey refer to the Anatolian village when talking about Turkish immigrants in Germany. Ertan was born and raised in Schöneberg, a middle-class neighbourhood in West Berlin. His grandfather was part of the first-generation of Turkish labour migrants (*Gastarbeiter*) who came to Germany in the 1960s. Ertan is 35 years old and has worked as an undertaker for six years. I ask him about the bureaucratic procedures involved in the steps leading up to a burial and he gives me a rundown of all the different government offices that he must visit to acquire the necessary papers and permits:

> I go to the *Bürgeramt* [civil office] and do an *Abmeldung* [unregister] and then I have to go to the *Standesamt* [registry office] and get the death certificate. Then I have to go to the *Gesundheitsamt* [health department] and get a health report, then go back to the *Bürgeramt* to get the *Leichenpass* [transit permit for a corpse] and then to the Consulate.

He underscores the fact that these operations take a lot of time, noting that his customers are unaware of the complexity of the procedures involved. Like Mert and Bora, he interprets their position as reflecting a rural mindset and makes no attempt to hide his sarcasm as he tells me

> Of course, since the customer knows everything, they say, when you pick up the corpse, let's just fax the paper to the *Standesamt*. I tell them, it's not that simple. I have to go to five different government bureaus first. When they hear that they are really surprised. Our people think that when there's a funeral, you can bury it within two hours or whatever, just like in the village. But this is Germany! There are bureaucratic procedures we have to do. But our people don't know this. Or they know it and don't want to admit it. And because of these situations, we are under a lot of stress. Our work isn't easy.

Although Ertan has no first-hand experience with rural funerary traditions, he uses the notion of the Anatolian village as a strategy to distinguish his own structural position from that of his customers. For all three undertakers quoted above, bureaucratic competence is a mark of distinction that delineates social integration. Germany is described as a highly regulated and bureaucratized country and one's ability to comprehend and navigate its bureaucracy is evidence of proper integration. In making the claim that knowledge of the legal-rational order is an important metric by which to judge social status, the undertakers draw on existing public discourses in Germany that take immigrants, particularly Muslims, as objects of pedagogical intervention by the state. In recent years the German government has launched a number of programmes aimed at creating institutional spaces for inter-religious dialogue, most notably the *Deutsche Islam Konferenz* (German Islam Conference). Such programmes have come under scrutiny for flattening differences across Muslim communities, creating divisions within German society, and for treating Muslims as a group that is to be governed and governable (see Peter 2010, Dornhof 2012).

While socio-political integration of immigrant communities has been a stated policy of the German state, in the examples presented above, the agents of change are not state actors, but private individuals from the immigrant community. What endows them with credibility to make judgements and interventions on topics beyond their immediate area of expertise – the burial of the dead – is linked to their own ability to move between the bureaucratic spaces of the German legal system and the cultural expectations of their customers. As Bora tells me later in our interview, 'I grew up between two different cultures but I must have picked up a lot of German traits. I work with appointments. I explain the rules to people. I tell them how the system works. I know very well that other [funeral] companies tell their clients

> 'Don't worry, we can ship the body tomorrow'. But it's a lie. I explain all the procedures to my customers. I tell them about the different governmental offices that I need to go to, the different paperwork that is required. I tell them this so that they have the information. But when I

> tell them this, I'm the bad guy. They'd rather hear 'don't worry, I'll take care of it'. Even if it's impossible. I don't like to do that and for that reason I guess I've become more German when it comes to these matters. That's how I am.

For Bora, educating his customers in the particularities of the regulatory structures governing the handling and disposal of the dead is a crucial, though risky, part of his job. While he admonishes his competitors for providing false information to bereaved clients, he also recognizes that his inability to meet customer expectations can hurt his reputation in the eyes of his clients. Nonetheless, his insistence on transparency and full disclosure demonstrates how an important aspect of his occupational identity is premised on cultural and political mediation. While his knowledge of the bureaucracy helps constitute his expertise, his ability to help guide families through an unfamiliar bureaucratic terrain is central to his work as a mediator between civil society and the state.

Alongside the immigrant communities that Berlin's Muslim undertakers serve, they also work hand-in-hand with various agencies and agents of the German state. Another dimension of the mediating work that they perform involves combatting negative stereotypes about Muslims in Germany. In the next section, I will analyse some of the strategies they use to establish their credibility as cultural representatives. By presenting themselves as responsible, professional, well-integrated, and knowledgeable individuals, the Muslim undertakers of Berlin attempt to demystify popular assumptions and misconceptions about Muslim immigrants in Germany. In some cases they take on the role of a spokesman and pedagogue willingly but in others they are put in a position where they are compelled to speak on behalf of others.

Countering stereotypes

Previous studies of death workers in Western Europe and North America have argued that undertakers often develop techniques to curb negative perceptions of the funeral industry and to reduce personal stigmatization. In his ethnographic study of funeral directors in four American states, Thompson (1991) claims that death workers use a variety of role-distancing techniques to combat stigmatization, including emotional detachment, humour, and countering stereotypes. Likewise, Howarth (1996, 88–91) suggests that undertakers adopt different strategies to allow them to 'pass as just another ordinary human being', including habitually distorting, omitting information, or evading questions about their occupation. In my own research, I found that stigmatization was not a major issue for the Muslim undertakers of Berlin, at least in terms of the stigma attached to their professional occupation. A larger and more politically salient problem that they faced in the course of their work was the stigma related to popular perceptions of Muslims in Germany. Consequently, one of the important tasks that they saw themselves performing was combatting negative ideas about Islam by countering stereotypes and prejudices through their own behaviour.

'I teach a lot of classes in hospitals and police stations about the things that people should pay attention to when there is a Muslim funeral' explains Ismail. He is a clean shaven, smartly dressed man in his early 50s, wearing a neatly pressed suit and tie. Ismail migrated to Germany at the age of six with his parents and was the first in his family to earn a university degree. Our interview takes place in his office, a light-filled building with its own morgue, sitting room, and garden. He has worked in the funeral business for seven years, having started as employee of the German firm *Ahorn-Grieneisen*. In 2011 Ismail left *Grieneisen* to start his own company, which he runs to this day. We have been talking about the Islamic funeral industry in Berlin and Ismail has spent the last few minutes chiding his competitors, whom he views as unprofessional and inexperienced. Unprompted, he switches gears to tell me about his efforts at educational outreach. 'Since 9/11' he continues,

> people in Germany get a little uneasy when they hear the word Muslim. I try to alleviate those fears in my classes. I often invite people to come visit my business because it's much easier to allay their concerns when they come in and see things for themselves.

In his classes, Ismail covers topics related to the handling of Muslim corpses, offers advice on how to treat dying Muslims and on steps that can be taken to establish bonds of trust with their families. 'Germans are really afraid of Muslims', he tells me.

> And after those events in the US they are even more afraid. When someone says 'Bismillahirrahmanirrahim' (In the name of Allah), the Germans will look around and say, 'What's going on? Is there a bomb? […] In the courses I teach I try to take away this fear.

For Ismail, negative stereotypes about Muslims in Germany are pervasive and have been heightened in the post-9/11 era. The conflation of Islam with violence is a symptom of a broader problem of misrepresentation and prejudice. Although his account might be slightly embellished, it is clear that Ismail sees a need to correct unfavourable images of Muslim immigrants by educating those who have regular contact with Muslims in their line of work. One strategy that he employs is explicitly pedagogical. By visiting hospitals and speaking with staff members his goal is to educate them on Islamic death rites and rituals so that they can provide proper care. Another strategy has to do with his own appearance and self-presentation.

> Most people who visit my funeral home expect to see a bearded man, a *hacı hoca* [Turkish slang for an ostentatious religious figure]. When they see me they are surprised. They ask me, "wait, are you a Muslim?" because they were expecting someone with a big beard [laughs].

Ismail expresses a certain pleasure in this sort of misrecognition. By not conforming to the expected image of a Muslim undertaker he challenges preconceptions about what a Muslim should look like. In embodying and presenting an alternative Islamic identity, Ismail hopes to dispel some of the myths that circulate in the German public sphere. As an undertaker, he does not represent any particular group or community. Yet as these examples demonstrate, Ismail willingly embraces the role of a public figure with a political mission. In challenging expectations about Muslims in Germany, Ismail provides a type of corrective cultural mediation.

Appearances can be deceiving however, and individuals who bear certain physical signs of a purported Islamic identity can face a different set of challenges related to racial profiling and discrimination. On a sunny afternoon I accompany Ertan to the *Landschaftsfriedhof Gatow*, a cemetery in the neighbourhood of Spandau, on the Western outskirts of Berlin. *Gatow* is one of the two cemeteries in the city with dedicated sections reserved for Islamic graves. We are greeted by two gravediggers and a cemetery administrator whom Ertan has known for many years. The four men have a cordial relationship and make frequent jokes with one another. As we approach them, they point to Ertan and exclaim 'Taliban! Look out! The Taliban is here!' Although it is all in jest, Ertan has frequently experienced such taunts in his private and professional life. He sees it as a challenge that he must personally overcome in order to correct misconceptions about Muslims in Germany. With his long beard, he knows that he might appear threatening to some people but attempts to counter the stereotype of the violent, fundamentalist Muslim through his personal interactions with civil servants and public officials.

> When I go to municipalities in the East, places like Pankow, Hellersdorf [neighborhoods in what was formerly East Berlin], people look at me and size me up. Dark hair, beard, Turkish, foreigner. When they see the beard they think Muslim

he explains, highlighting the link between physical appearance and presumed religiosity.

> When they make that connection it's over. Maybe they imagine Osama Bin Laden, or a bomb, or the twin towers. But when I start speaking to them in German, they are really surprised … I can sense a change in their tone of voice. And maybe because of this, I'm able to give them a different example of what a Muslim or a Turk looks like.

As mentioned above, Ertan was born and raised in Berlin. He has native fluency in German. Yet because of the way he looks, people make certain assumptions about him, concluding that he is unlikely to speak proper German. Language is an intrinsic part of social identity and questions of linguistic competence are particularly salient in debates around immigration and integration in Germany. With the reform of Germany's citizenship laws in 2000, proficiency in the German language was established as a precondition for naturalization and the acquisition of German citizenship (Piller 2001). This legislation reflects the popular perception that immigrants in Germany lack the requisite language skills to fully participate in German society. Given the tenor of these debates, it is not surprising that Ertan encounters some degree of disbelief when he is able to communicate clearly and effectively. By doing so, he is able to challenge some of the misconceptions about immigrants' linguistic capabilities.

Alongside his efforts to correct stereotypes about language, Ertan is often compelled to speak on behalf of Muslims or Turkish immigrants in Germany. Unlike Ismail, who seeks venues to speak to public officials on topics related to Islam, Ertan's interventions and mediations occur during routine visits to bureaucratic offices during the course of his work. During our interview I ask him about the funeral company that has been accused of selling passports and whether it has impacted his own business in any way. He explains that he faces heightened suspicion in the municipal offices and interprets this as part of a broader pattern of discrimination. 'If one person makes a mistake' he tells me, 'we all suffer for it'. Ertan questions whether German funeral companies have had to face a similar degree of scrutiny in the aftermath of the passport scandal, and recounts how he has had to explain to numerous civil servants in the municipal offices that he has no connection to the company that is under investigation. 'These are people I've known and worked with for years', he continues, in reference to the civil servants whose offices he regularly visits to file paperwork.

> Now they ask me questions like, "Why are Turks so angry? Does Islam allow that?" And I tell them this has nothing to do with Islam, it has to do with the person. Being hot tempered is a personality trait!

The erasure of difference and the homogenization of diversity is one of the pernicious effects of stereotyping. Ertan's comments draw from his own personal experiences but reflect a broader practice of taking individual behaviour as indicative of an entire group. Although as an undertaker, Ertan would not be expected to weigh in on theological issues or to provide sociological analyses of group dynamics, as someone who is read as a Muslim he frequently finds himself in a position where he is required to do so. This suggests that a certain type of cultural and political mediation is characteristic of professionals like the Muslim undertakers of Berlin, who operate between civil society and the state.

More broadly, it points to an important feature of the lived experience of Western Muslims in Europe and North America today. To a certain extent, there is a constant demand placed on Western Muslims to speak for, and on behalf of Islam or the ethnic or national communities, they are perceived as being a part of. Such demands not only essentialize minorities by foregrounding certain aspects of their identity, but also place an undue burden on individuals belonging to minority groups to speak on behalf of the entire community. As Norton (2013, 42) has argued, the compulsory speech acts required of Muslims in the West attest to the 'radical narrowing of the right to free speech'. While Ertan and Ismail's efforts to counter existing stereotypes about German Muslims can be seen as small steps in bringing about broader shifts in public perceptions, it is important not to overlook the asymmetrical dynamics of power at work in such interactions. The demand that all Muslims be prepared to speak on behalf of Islam threatens not only to trivialize politics, but to strengthen the divisive binaries that posit a hierarchy of citizenship amongst those who belong and those who do not in contemporary Germany.

Conclusion

This article has sought to highlight the work of political and cultural mediation that is performed by the Muslim undertakers of Berlin. As intermediaries between immigrant communities and the German state, undertakers help families navigate the cultural, religious, political, and legal landscapes that structure the transitions from life to death. Their cultural capital and professional credibility is derived from their ability to anticipate and manage the expectations of their customers while guiding them through the German bureaucracy. Weber (1978, 225) famously asserted that 'bureaucratic administration is domination through knowledge'. As this article has shown, the authority of the Muslim undertaker is in part a function of his knowledge of the bureaucracy.

In mediating between civil society and the state, the Muslim undertakers of Berlin not only help instruct immigrant families in the legal-rational order of the German bureaucracy, but also engage with members of that order to counter and dispel stereotypes about Muslims and Islam in Germany. Their role as a spokesperson is at times taken up willingly, but can also be thrust upon them. Consequently, their ability to serve as cultural translators or political brokers can be seen both as a positive effort to fight prejudice and as an example of the uneven power dynamics that frame contemporary discussions about Islam in the West.

Recent scholarship has stressed the need for the provision of culturally appropriate palliative and end-of-life care in places such as hospitals and hospices, while insisting that practitioners remain mindful of the diversity of lived experiences that exist as much within cultures as across them (Gunaratnam 2013). This article has attempted to demonstrate that the intercultural negotiations around death and dying do not conclude with the death of an immigrant. Post-mortem procedures are governed by a different set of rules and regulations that raise culturally inflected questions about the proper treatment and handling of corpses. In migratory settings, undertakers play an important role in mediating between the expectations of their customers and the laws of the state. As such, they are not simply professionals that oversee the burial of the dead, but cultural and political mediators that preside over pedagogical moments in the transitions from life to death.

Note

1. A 'person with a migration background' is an official category employed by the Federal Statistics Office that includes everyone who migrated to the Federal Republic of Germany after 1949 as well as all foreigners born in Germany after 1949, and all Germans born in Germany with one parent who immigrated to Germany after 1949, or one parent who was born as a non-German citizen in Germany.

Acknowledgements

Earlier versions of this article were presented at the International Migration, Integration, and Social Cohesion conference and at the annual meeting of the Western Political Science Association. My thanks to Matthew Berkman, Guzmán Castro, Orfeo Fioretos, Kambiz GhaneaBassiri, Justin Gest, Jeff Green, Matthew Handelman, Danielle Hanley, Alistair Hunter, Anne Norton, Thea Riofrancos, Eva Soom, Bob Vitalis, and two anonymous reviewers for their thoughtful comments and criticisms. Though I cannot acknowledge them by name, I am grateful to the undertakers quoted herein. This study would not have been possible without their candour and enthusiasm.

References

Akyel, D., 2013. Qualification under moral constraints: the funeral purchase as a problem of valuation. *In:* J. Beckert and C. Musselin, eds. *Constructing quality: the classification of goods in markets.* Oxford: Oxford University Press, 223–246.

Attias-Donfut, C. Wolff, F.C., and Dutreuilih, C., 2005. The preferred burial location of persons born outside France. *Population-E*, 60 (5–6), 699–720.

Balkan, O., 2015a. Burial and belonging. *Studies in ethnicity and nationalism*, 15 (1), 120–134.

Balkan, O., 2015b. Till death do us depart: repatriation, burial, and the necropolitical work of Turkish funeral funds in Germany. *In:* Y. Suleiman, ed. *Muslims in the UK and Europe.* Cambridge: Cambridge University Press, 19–28.

Bremborg, A., 2006. Professionalization without dead bodies: the case of Swedish funeral directors. *Mortality*, 11 (3), 270–285.

Bundesverband Deutscher Bestatter e.V., 2015. Häufig gestellte Fragen – Allgemein [online]. Available from: https://www.bestatter.de/meta/news-termine-presse/haeufig-gestellte-fragen-allgemein/ [Accessed 19 May 2015].

Cahill, S., 1995. Some rhetorical directions of funeral direction: historical entanglements and contemporary dilemmas. *Work and occupations*, 22 (2), 115–136.
Carpenter, B., *et al.*, 2015. Scrutinising the other: incapacity, suspicion, and manipulation in a death investigation. *Journal of intercultural studies*, 36 (2), 113–128.
Dornhof, S., 2012. Rationalities of dialogue. *Current sociology*, 60 (3), 382–398.
Doughty, C., 2014. *Smoke gets in your eyes and other lessons from the crematorium*. New York: W. W. Norton & Company.
Félix, A., 2011. Posthumous transnationalism: postmortem repatriation from the United States to Mexico. *Latin American research review*, 46 (3), 157–179.
Gardner, K., 2002. Death of a migrant: transnational death rituals and gender among British sylhetis. *Global networks*, 2 (3), 191–204.
Gunaratnam, Y., 2013. *Death and the Migrant: bodies, borders, and care*. London: Bloomsbury.
Howarth, G., 1996. *Last Rites: the work of the modern funeral director*. Amityville: Baywood.
Initiative Kabir, 2015. Muslimische Bestattungskultur und Grabfelder in Deutschland [online]. Available from: http://www.initiative-kabir.de [Accessed 15 May 2015].
Jassal, L., 2014. Necromobilities: the multi-sited geographies of death and disposal in a mobile world. *Mobilities*, 10 (3), 1–24.
Jonker, G., 1996. The knife's edge: Muslim burial in the diaspora. *Mortality*, 1 (1), 27–43.
Koopmans, R., 2004. Migrant mobilisation and political opportunities: variation among German cities and a comparison with the United Kingdom and the Netherlands. *Journal of ethnic and migration studies*, 30 (3), 449–470.
Kselman, T., 1993. *Death and the Afterlife in modern France*. Princeton, NJ: Princeton University Press.
Laderman, G., 2003. *Rest in peace: a cultural history of death and the funeral home in 20th century America*. Oxford: Oxford University Press.
Lynch, T., 1997. *The undertaking: life studies from the dismal trade*. New York: W.W. Norton & Company.
Marjavaara, R., 2012. The final trip: post-mortal mobility in Sweden. *Mortality*, 17 (3), 256–275.
Mazzucato, V., Kabki, M., and Smith, L., 2006. Transnational migration and the economy of funerals: changing practices in Ghana. *Development and change*, 37 (5), 1047–1072.
New York Times, 2014. German funeral homes sold passports of the dead [online]. Available from: http://www.nytimes.com/aponline/2014/05/15/world/europe/ap-eu-germany-human-trafficking.html [Accessed 20 May 2015].
Norton, A., 2013. *On the Muslim question*. Princeton, NJ: Princeton University Press.
Oliver, C., 2004. Cultural influence in migrants' negotiation of death. The case of retired migrants in Spain. *Mortality*, 9 (3), 235–254.
Parsons, B., 1999. Yesterday, today, and tomorrow. The lifecycle of the UK funeral industry. *Mortality*, 4 (2), 127–145.
Peter, F., 2010. Welcoming Muslims into the nation: tolerance, politics, and integration in Germany. *In:* J. Cesari, ed. *Muslims in the west after 9/11: religion, politics, and law*. London: Routledge Press, 119–144.
Piller, I., 2001. Naturalization language testing and its basis in ideologies of national identity and citizenship. *International journal of bilingualism*, 5 (3), 259–277.
Prendergast, D., Hockey, J., and Kellaher, L., 2006. Blowing in the Wind? Identity, materiality, and the destination of human ashes. *The journal of the royal anthropological institute*, 12 (4), 881–898.
Pringle, R. and Alley, J., 1995. Gender and the funeral industry: the work of citizenship. *Journal of sociology*, 31 (2), 107–121.
Renteln, A., 2001. The rights of the dead: autopsies and corpse mismanagement in multicultural societies. *South atlantic quarterly*, 100 (4), 1005–1027.
Schulz, F., 2013. The disappearing gravestone: changes in the modern German sepulchral landscape. *In:* M. Aaron, ed. *Envisaging death: visual culture and dying*. Cambridge: Cambridge Scholars, 10–26.

Smith, S., 2010. *To serve the living: funeral directors and the African American way of death.* Cambridge: Harvard University Press.
Statistik Berlin Brandenburg, 2014. Statistische jahrbuck Berlin [online]. Available from: https://www.statistik-berlin-brandenburg.de/produkte/produkte_jahrbuch.asp [Accessed 15 May 2015].
Thompson, W., 1991. Handling the stigma of handling the dead: morticians and funeral directors. *Deviant behavior*, 12 (4), 403–429.
Venhorst, C., 2012. Islamic death rituals in a small town. *Omega*, 65 (1), 1–10.
Walter, T., 2005. Three ways to arrange a funeral: mortuary variation in the modern west. *Mortality*, 10 (3), 173–192.
Weber, M., 1978. *Economy and society.* Berkeley: University of California Press.
Zirh, B., 2012. Following the dead beyond the 'nation': a map for transnational Alevi funerary routes from Europe to Turkey. *Ethnic and racial studies*, 35 (10), 1758–1754.

The Importance of a Religious Funeral Ceremony Among Turkish Migrants and Their Descendants in Germany: What Role do Socio-demographic Characteristics Play?

Nadja Milewski and Danny Otto

Faculty of Economic and Social Sciences, University of Rostock, Rostock, Germany

ABSTRACT

Our paper analyses the attitudes of Turkish migrants and their descendants in Germany regarding the importance of a religious funeral ceremony. Previous research provides competing hypotheses on the intergenerational transmission of religiosity in migrant communities, such as, declines in religiosity due to assimilation versus maintenance of religiosity as a means to ethnic identity formation. Quantitative research however has not yet considered funerals. Our study utilises data from the Generations and Gender Survey; our sample comprises roughly 4000 people of Turkish migrant background aged 18–81, most of whom are Muslims. We apply logistic regression methods to attitudes regarding the importance of a religious funeral ceremony. More than 80 per cent of the respondents maintained that a religious funeral ceremony was important. Examination of individual characteristics revealed variation by education, partner's origin, and citizenship. Overall, however, Muslim funeral traditions are sustained across first- and second-generations.

This study examines the importance Turkish migrants and their descendants living in Germany ascribe to a religious funeral ceremony. Emigrants from Turkey are not only the most numerous group of third-country nationals living in Europe today, they have also settled in more European countries than any other national-origin group. Since the 1960s, this community has been living and aging in Western European destination countries, and subsequent generations have been born and grown up in Europe. The majority of Turkish migrants[1] and their descendants adhere to Muslim religious practices, whereas the non-migrant populations primarily belong to Christian denominations or have no religious affiliation.

Over the past two decades, scholars from various disciplines have been drawing attention to both the aging process and health care provision at the end of life as well as the religiosity of immigrants in these contexts. These topics however appear separately in the literature. Death in migration contexts has up until recently been an 'under-investigated grey zone' (Zirh 2012, p. 1759). Previous studies on this subject come mainly from social or cultural anthropology, history, or human geography. They explain funerary practices, such as the rituals that surround caring, dying, mourning, and burying as expressions of feelings of group belonging and identity (Reimers 1999, Ansari 2007), of place attachment (Balkan 2015a), as well as of religious acculturation and immigrant integration (Jonker 1996). Funerary practices have been tied to transnational engagements among migrant communities, institutions, and policy actors (Zirh 2012, Balkan 2015b). These works employed qualitative research to explain the meaning and development of attitudes and practices concerning end-of-life rituals.

Parallel to this, there has been an ongoing debate on the development of religiosity among immigrants in Western Europe. Previous research, particularly from the sociological, socio-demographic, socio-psychological, and political sciences, has brought forward competing hypotheses on the development of religiosity and its impact on other domains of structural integration of immigrants and their descendants (Foner and Alba 2008). Religiosity may be maintained over generations due to religion being a means of group identity and/or an expression of intergenerational transmission; against this view, religiosity may decline due to various processes of adaptation to the host populations (see, Phalet *et al.* 2008, Diehl and Koenig 2009, Van Tubergen and Sindradóttir 2011). The focus has been mainly on Muslim immigrants in countries with Christian traditions. Religious funeral practices have not yet received attention in this strand of research, however.

This paper aims to bridge these two research fields. It adds a quantitative perspective by studying the role of individual characteristics in shaping attitudes about the importance of a religious funeral ceremony. We used representative survey data from Germany and asked about the extent to which Turkish migrants and their descendants prefer a Muslim funeral ceremony and whether these attitudes remained stable or varied between generations. Our paper is organised as follows: the review of the literature opens with statistical information about Turks in Germany. We then summarise qualitative studies on funeral practices of international migrants, before outlining some theories and findings on the religiosity of immigrants and their descendants. Our paper employs data from the Generations and Gender Survey (GGS) (2005/2006), and studies the role of classic socio-demographic indicators, such as sex, age, education, and nationality, on the self-reported importance of a religious funeral ceremony. We show that compared to other religious indicators, funeral ceremonies are seen as the most important religious practice, even among those respondents who did not attribute high importance to religious practices in other life stages. We also demonstrate that higher education, German citizenship, and a union with a non-migrant partner, typically used as measures for integration and adaptation processes, decreases religiosity. Overall, however, there is a high retention of Muslim funeral traditions over generations in immigrant communities.

Aging immigrants, religiosity, and funerary practice

Before turning to religiosity regarding funerals, we provide some statistical information about the immigrant population in Germany, being one of the largest immigrant populations in Western Europe. Today, about every fifth person in a population of roughly 80 million either migrated themselves to Germany, or were born to one or two parents who were born abroad (Destatis 2014). The age structures of non-migrant and immigrant populations differ: the non-migrant population has been declining and aging due to persistent fertility levels below 1.5 (which means that a given birth cohort is not fully replaced by the following cohort) and due to increasing life expectancy. The immigrant population is still younger on average than the non-migrant population, because large-scale immigration began only in the 1950s, and the immigrants' fertility levels were higher on average. Hence, the share of the German population with an immigrant background has been rising steadily and will continue to do so: today, the share with an immigrant background in the age group over 85, for example, is only about 6 per cent. Among those who are 50–60 years old today (and will be 80–90 years old in 30 years) the respective percentage is about 16 per cent, and among children who are age 10 and younger today it is about 35 per cent (Destatis 2014). Hence, Germany is becoming increasingly heterogeneous with regard to demographic behaviour (fertility and mortality) as well as religious and cultural practices of the respective population groups (Swiazny and Milewski 2012). The question of funerary practices will therefore gain importance – perhaps not only, but mainly – because of demographic developments: the absolute number of deaths within the immigrant population will increase over the coming decades.

Turks are the largest immigrant group from a single country, and they are the biggest Muslim community in Germany. Immigration from Turkey started in 1961 with labour recruitment and later continued mainly as family chain migration (Abadan-Unat 1995). Family formation and family reunion are still the largest flows of immigration from Turkey today, whereas the numbers of refugees and students are smaller. In total, about 2.5 million persons in Germany have a Turkish migrant background. Of these, about 1.5 million migrated themselves from Turkey (BAMF 2011). Today, migrants' families in Germany are witnessing the arrival of third- and fourth-generations. The processes of incorporating these migrant-origin communities are rather complex because there is always a new first-generation, as immigration from Turkey continues and transnational relations between Germany and Turkey persist (Pries 2010). For example, transnational partner choice is demonstrably higher among Turks than in other immigrant groups, and – despite a decline – is still rather high in the second-generation (Hamel *et al.* 2012). Among elderly migrants, transnational mobility between Germany and Turkey has become a lifestyle, developing from life-long migration biographies that left the decision open whether the migrant would eventually return to his/her country of origin (Krumme 2004). Throughout the life course, transnational mobility and family networks are highly intertwined. Intergenerational relationships are relatively strong in Turkey, and they even become stronger in a migratory setting (Baykara-Krumme 2013). In crucial life-course transitions, such as partner choice and union formation, the involvement of the family has remained an expression of intergenerational relationships and social cohesion within a migrant community, which adapts its rituals to a transnational context (Aybek *et al.* 2015; Milewski and Huschek 2015). This extends up to and after the final phase of life, as recent research on Turks in Germany has shown. Strumpen (2012) found that mourning practices are perceived as an obligation and therefore another reason for transnational commuting, for example, to visit the grave of one's ancestors.

Religion provides both a cultural framework to make sense of one's own life and death as well as psychological resources and coping strategies (Türkis and Seeberger 2012). Therefore, even though religiosity may decline overall due to trends of secularisation, religious traditions continue to guide funerary practices:

> [...] even if single individuals do not regard themselves as religious, they are still able to employ religious rites in meaningful ways. This is because rituals in a non-verbal language convey to people where they belong. By making choices on how to announce the death and the funeral, the place for the funeral, officiating official, the ritual service, flower tribute, coffin, music, gravestone, and so forth, the bereaved communicate not only who the deceased was but also who they are and where they belong. (Reimers 1999, p. 162)

Furthermore, the significance of rituals surrounding dying, burial and mourning may not only be religious; rather, in a migratory context, funerary rites can also be a link to a social/ethnic group or to a place/country: '[...] the issue of death throws into relief the most important cultural values by which people live their lives and evaluate their experiences. Life becomes transparent against the background of death, and fundamental social and cultural issues are revealed' (Metcalf and Huntington 1991, p. 24–5). Conceptions of individual and collective identities may be challenged (see, Berger and Luckmann 1979).

For Muslims in Germany, the majority of whom are Sunni from Turkey, the topic of death in a migration and multi-religious context has received moderate scholarly attention since the 1990s (Höpp and Jonker 1996, Tan 1998, Kaplan 2004, Kuhnen 2009). Unfortunately, except for Jonker (1996), this work has received little recognition. As the Turkish population ages, the number of deaths has been increasing. In 2010 there were 4150 deaths of Turks in Germany (Destatis 2012), but no official statistics exist on where these persons were buried. The little quantitative information available indicates that the vast majority of Muslims who lived in Germany returned to their countries of origin: Jonker (1996), e.g., estimated that only about two per cent of Berlin's Muslim community was actually buried there; and this situation has changed only slightly since then. The majority returns to their country of origin posthumously; estimations vary between 70 and 95 per cent (Zirh 2012, Balkan 2015a, 2015b). A high rate of post-mortem repatriation is found for other Muslim groups in other Western European countries (Zirh 2012). The place of the funeral weighs heavily in conceptions based on belonging to an ethnic and/or religious group (Balkan 2015a).

For Muslims, scriptural stipulations on place of burial propose that a deceased person should be buried where he/she had lived (Kuhnen 2009): the transportation of a corpse to another site for burial is however allowed under certain conditions (Campo 2006). The possibility of post-mortem mobility has thus been a central question in work on funerary practices among Turks or other Muslim migrant populations in Europe (Zirh 2012, Balkan 2015a). Previous research has linked the high share of post-mortem expatriations to various explanations.

Some explanations centre on legal conditions. Despite variation among traditions, the Islamic legal canon rules pertaining to Muslim funerals range from 'prescribed' to 'recommended' then to 'forbidden' practices. Prescriptions include ritual bathing of the body, shrouding, and funeral prayers. The corpse should be turned on its right side, facing Mecca. In contradiction with German law, Islamic law stipulates that the burial should take place without delay and coffins should not be used (Campo 2006, Balkan 2015a).

Legal conditions apply at a national level, supporting a rough dichotomisation of cultural traditions by country of origin, even though there is variation within these countries as well as conflicts between the various schools of Islamic jurisprudence (Zirh 2012). The Turkish Funeral Funds in Germany, in which many of the Turkish migrants are members, support repatriation by providing material incentives for funerals in Turkey and disincentives for burial in Germany (Karakasoglu 1996, Türkis and Seeberger 2012, Balkan 2015b). Through such institutions, an image is promoted within the community that it is only in Turkey that migrants can fully adhere to cultural tradition and have a 'pure' funeral. This image is based on the distinction between being Turkish/Muslim/strangers versus Germans/Christians/natives (Kuhnen 2009, Strumpen 2012) and is also linked to experiences of social exclusion and discrimination in daily life (Balkan 2015a).

In addition to legal conditions, other explanations for post-mortem expatriation pertain to transnational family and community networks in which biographies are segmented and identities are fluid. Many Turks who came to Germany as guest workers had originally intended to return to Turkey. This envisaged return may have ended up being an illusion, but it nonetheless served to protect people from the image of dying as a stranger abroad. Post-mortem migration to Turkey may thus be perceived as the fulfilment of the migration project – if only after the end of life (Tan 1998). The choice of one's funeral place is to be negotiated between migrant generations. Whereas return intentions may still exist among the first-generation, subsequent generations have become more connected and integrated in Germany. A visit to their (grand)parents' graves would be facilitated if they were buried in Germany. Hence, the comprehension that the subsequent generations of the family will live in Germany may cause Turkish migrants and/or their descendants to choose their place of burial in Germany (Türkis *et al.* 2012, Balkan 2015a).

The understanding that a share of Muslims and converted persons may prefer to be buried in Germany has initiated change in Germany's funerary geography. More than 250 cemeteries provide sections for Islamic funerals (Balkan 2015a). German officialdom seems able to adapt to the new clientele, and so do the bereaved, the Muslim institutions and wider communities – at least to a certain extent and with variation by country of origin. They have found ways to deal with ambivalence and contradictions between Muslim traditions and German laws and rules (Tan 1998, Türkis and Seeberger 2012). Jonker (1996) identified two coping strategies among religious leaders:

> Some religious leaders will make room for new expressions, thus widening the possibilities of individual and collective readjusting. Others, wishing to protect their community from foreign influences, try to denude the burial ritual of all non-Islamic elements, thus forbidding symbolic attempts to bridge the gap between one's actual life and one's past. (Jonker 1996, p. 27)

Often it is the undertaker's task to find pragmatic solutions between bureaucratic demands and theological considerations. Such adaptations include not only the preparation of the body and the gravesite, but also the funeral ceremony and other traditions, for example, taking films or photos of the deceased in order to maintain communication between the persons at both ends of the migratory chain (Jonker 1996).

Although these fieldwork studies demonstrate the complexity of (religious) funerals in transnational networks, quantitative studies on the religiosity of Turks in Germany as well as on other immigrant groups in Western Europe have not included funerary practices or attitudes towards the end of life. They have, however, paid increasing attention to within-group comparisons by generation. The indicators most often studied have been religious affiliation, attendance at religious meetings, and/or the degree of subjective religiosity.

Turkish immigrants in Germany have been shown to be more religious on average than non-migrants, which has been mainly attributed to the overall higher degree of religiosity in Turkey (Diehl and Koenig 2009). Comparing the generations to each other, however, the empirical evidence does not produce a clear pattern of adaptation to non-migrant attitudes and behaviours over generations, as was suggested by classical assimilation theory (Alba and Nee 2003) or by the secularisation paradigm (see,. Phalet *et al.* 2008). Instead, Diehl and Koenig (2009) found that religious attitudes and behaviours of Turks in Germany are a mixture of different processes: a decrease in religiosity over generations may be caused by a higher degree of structural integration, mainly measured as higher education and occupation (see, Van Tubergen 2006). Stability or even an increase in religiosity may be caused by a shift in meaning of religious practices towards a symbolic religion (see, Gans 1979) and/or strong intergenerational transmission and solidarity in immigrant families (see, Phalet and Schönpflug 2001, Baykara-Krumme 2013). Similarly for Turks in the Netherlands, Phalet *et al.* (2008) suggested a selective secularisation conditional on social and structural incorporation. We will come back to this after the description of the data we used.

Data and variables on religion and religiosity

When we take stock of the existing quantitative studies on religiosity and religious funerals among international migrants, the yield is quite small. Very few surveys in single countries explicitly include immigrants or focus solely on immigrant groups of a certain origin. On a country-comparative level, several surveys have been carried out that included international migrants or made them the specific target population. Special attention has been given to Muslim immigrants from Turkey or Arabic-speaking countries living in Western Europe. These surveys and related quantitative studies also asked questions about religious membership and religiosity in general. Specific questions concentrated on items related to the five pillars of Muslim practice: the creed, the daily prayers, participation in Ramadan, the pilgrimage to Mecca, and donations to the poor. Other questions referred, for example, to the consumption of halal food, the wearing of a headscarf, or preference for religious education. All of these questions were, however, concerned solely with attitudes and practices during a person's lifetime (as in Phalet *et al.* 2008, Haug *et al.* 2009, Smits *et al.* 2010, Van Tubergen and Sindradóttir 2011, Phalet *et al.* 2012, El-Menouar 2014).

The only data set we are aware of that contains a question concerning funerals is the GGS.[2] The target population was persons aged 18–79 years. In Germany, the first wave of the main survey – all persons of the resident population who were linguistically able to follow the interview, regardless of their nationality – was undertaken in 2005. Germany was the only country in this international cooperation to explicitly examine its largest foreign nationality – migrants from Turkey – by including a specific subsample one year later (Ruckdeschel *et al.* 2006, Ette *et al.* 2007). Our study used respondents from both subsamples if they migrated from Turkey or if their mother or father was born in Turkey and migrated to Germany. With this strategy, our sample consisted of 4004 Turkish migrants and their descendants in total, of which about 7 per cent had German citizenship and 93 per cent Turkish nationality. The 1909 women and 2095 men were 18–81 years old when they participated in the survey.

The GGS contained one question related to religious affiliation and five questions regarding religiosity.[3] The majority of our sample was of Muslim affiliation (about 94 per cent). There were about two per cent each belonging to a Christian Church or to another denomination. Only a few declared no religious affiliation (about 1 per cent).

The crucial survey question for our study was the one on the importance of a religious funeral: 'It is important that a funeral also contains a religious ceremony' (translated from German). The answer categories ranged from 'strongly agree' to 'strongly disagree' and were provided as a five-point scale. We dichotomised these answers into important ('absolutely agree' and 'agree') and not important ('neither agree nor disagree', 'disagree', and 'absolutely disagree'). Although technically the survey question asked about 'religious' funerals, as the majority of the respondents were Muslims we assume that their answers can be interpreted as the importance attached to 'Muslim' funeral rites. In the total sample, the majority (over 80 per cent) of respondents perceived a religious funeral ceremony as important.

The GGS survey contained two more questions on the importance of religious rituals for crucial life-course events – marriage and baptism[4] – with possible answers similar to the question on funerals, ranging from strong agreement to absolute disagreement. Religious marriage rituals were declared important by about 70 per cent (we proceeded to use this variable similarly as for funerals). Another question regarding religiosity inquired into goals that are important for childrearing. Respondents could select three areas out of 11 which they viewed as important for childrearing, with religiosity being one. In our study, participants were considered to have religious childrearing goals if 'religious belief' was amongst the top-three areas selected. About 30 per cent of the total sample ascribed to religious aims in upbringing. Finally, one question asked about frequency of attendance at places of worship. The response categories, originally compiled with nine characteristics, were combined into three categories: more than once a week, once a week, and less than once a week.

Method and socio-demographic characteristics

The dependent variable in our study is the importance of a religious funeral ceremony. We applied logistic regression models and estimated the probability of considering religious funerals as important compared to the reference group, who were those persons who did not consider a religious funeral important (Larose 2006). We carried out multivariate models for women and men separately and included a number of socio-demographic control variables. The results are displayed in odds ratios. They denote the chances that a group is statistically different from the reference group in the respective variable, which is represented by the value 1.[5]

In addition to nationality and the parents' country of birth, the sample was differentiated by place of birth of the respondent and age at immigration, if the respondent was born in Turkey. The first immigrant generation was composed of participants who were born in Turkey and were 16 years or older when they immigrated to Germany. The members of the second generation were defined as being either born in Germany and having at least one Turkish parent, or being born in Turkey and having moved to Germany before the age of 16. The sample consisted of about 51 per cent first-generation migrants and 49 per cent belonging to the second-generation. In the literature, the definition of second-generation varies by research topic and discipline. As the second-generation born at destination in Europe is generally rather young, our sample also has a relatively young age structure. Hence, we chose to group those migrants who moved as children or during adolescence and those who were born in Germany together (see, Rumbaut 2004).

In previous studies, religious attitudes were shown to vary according to classic demographic variables such as sex and age, with women being more religious than men and religiosity increasing with age (see, Van Tubergen 2006). In our sample, the sex distribution was almost even (48 per cent women and 52 per cent men in total) and the mean age of the respondents was about 38 years. About three quarters of our sample were between 18 and 45 years old and only about 13 per cent (mostly first-generation migrants) were close to or already in retirement age (that is, 56 years and older).

As previous studies indicated a correlation between increased education and declining religiosity (see, Van Tubergen 2006, Maliepaard *et al.* 2010, Van Tubergen and Sindradóttir 2011), we included educational attainment in our analysis. About 43 per cent of our sample had obtained a primary level of formal education, and 17 per cent had finished lower secondary education. About 8 per cent left school with an upper secondary education degree, and about 3 per cent acquired tertiary education (as this last group is so small, we grouped upper secondary and tertiary education together). About 20 per cent had not attained any degree. A few respondents were still enrolled in education at the time of the survey (about 2 per cent) or achieved another school degree (5 per cent).

Concerning household composition, previous studies on religiosity differentiate between marital status and showed a higher degree of religiosity among individuals who are or were married as compared to never-married persons (see, Van Tubergen 2006, Maliepaard *et al.* 2010, Van Tubergen and Sindradóttir 2011). Regarding the background of the partner, Smits *et al.* (2010) showed that the presence of a co-ethnic partner increases religiosity compared to Muslim migrants whose partner had a different ethnic background. In our sample, we can trace whether the respondent shares her/his household with a partner, but further distinctions by marital status are not possible (Ette *et al.* 2007): however, the prevalence of non-marital cohabitation is rather low among the Turkish communities in Europe, even in the second-generation (Hamel *et al.* 2012). The information given in the survey allows us to account for the country background of the partner. The majority of respondents lived with a partner of Turkish origin; more than half of them lived with a partner who had migrated from Turkey (54 per cent), or with a partner from Germany who had a Turkish migrant background (9 per cent). A small group of respondents shared a household with a German non-migrant partner (5 per cent). Partners with other migrant backgrounds were rare (3 per cent). About 28 per cent had no partner in the household.[6]

Religious attitudes and practices among women and men of Turkish migrant background

We turn now to our results, starting with bivariate findings concerning religious attitudes by sex (Figure 1). A funeral was the most important religious practice, with

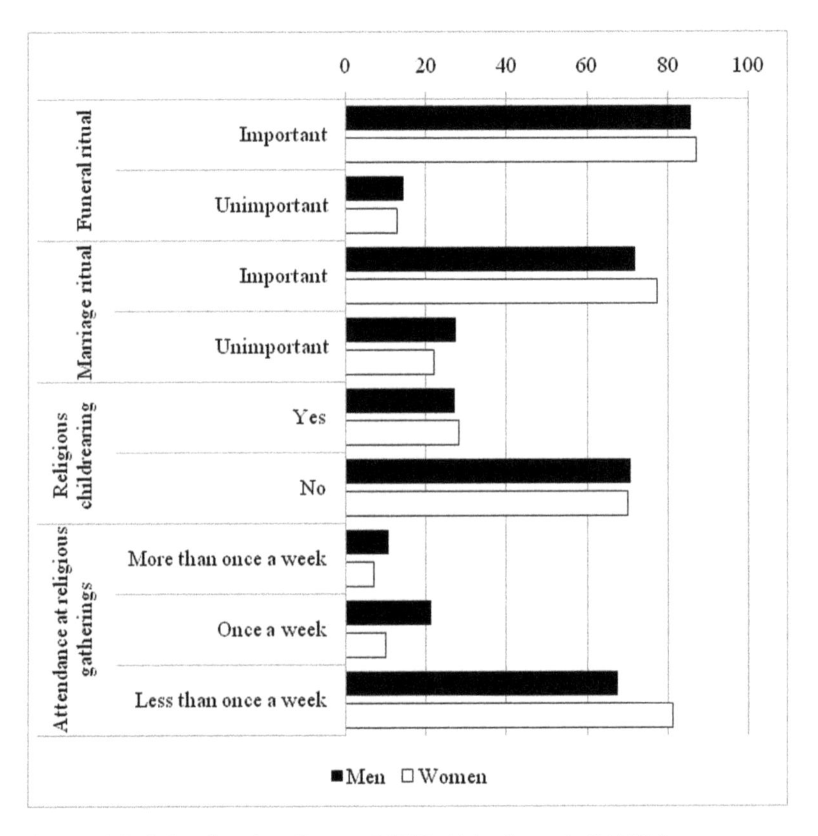

Source: Calculations based on German GGS/Turkish subsample (2005/06).
Note: N=4004; missing values not displayed.

Figure 1. Religious attitudes and practices among persons of Turkish migrant background in Germany, by sex (in per cent).

more than 80 per cent agreement. A religious marriage ritual received about 20 percentage points less. Roughly every fourth respondent listed religion as one of their three most important childrearing goals. About 20 per cent noted that they attended prayers once a week. About two thirds of the respondents, however, attended religious gatherings less often.

The answers did not vary much between the sexes, although there was a slight trend of women attributing more importance to the respective religious rituals in the life course. By contrast, we found that men attended religious services more often than women (see, Van Tubergen and Sindradóttir 2011). This may be traced back to gender-specific guidelines in Islam; that is, the category 'once a week' may mainly refer to the Friday congregational prayer, which men are required to attend whereas women are not obliged to do so (McAuliffe *et al.* 2002).

Comparing religious attitudes of the first- and second-generations, members of both attached importance to religious funeral ceremonies almost equally often (87 per cent and 85 per cent), whereas the importance of religious weddings was higher in the first than in the second (78 per cent and 71 per cent). This suggests first that Turkish migrants considered funerals to be the life course event where a religious or cultural tradition is most important. Second, these attitudes appear rather stable between the generations.

In our exploratory analyses, we carried out the same steps for Germans with no migrant background. They showed in each indicator a lower degree of religiosity. But even among the latter group, a religious funeral received the highest share of agreement; about 56 per cent ascribed the same level of importance to a religious funeral ceremony as those with a Turkish migrant background.

Individual characteristics influencing attitudes towards a religious funeral ceremony

We then investigated the effect of the socio-demographic indicators on the dependent variable, considering a religious funeral ceremony important. Figure 2 displays the results for women and men separately, because we found that the effect of the covariates varied partially between the sexes, that is, the direction and/or the size of their effect was different. Overall, our findings corresponded to those in the literature on other religious variables.

Women who belonged to the second-generation were about 10 per cent less likely to view a religious funeral as being important compared to women of the first-generation, men of the second-generation were 15 per cent more likely to do so than men of the first-generation (these differences were not statistically significant, though). In terms of age, we found that older respondents were more likely to consider a religious funeral important than younger Turks. These differences were significant for the oldest age group: men and women aged 65 and older considered a religious funeral more than twice as important compared with persons aged 26–35. This suggests that with increasing age and personal experience of loss, people's thoughts turn more and more to the idea of life's finiteness. It may also indicate that members of older birth cohorts were generally more religious than younger ones, in particular because many of the former migrant workers and their families originated from rural areas in Turkey, where religiosity may be higher than in cities (Türkis *et al.* 2012).

When we look at the indicators related to integration in the host society and to secularisation processes, we also find our results confirming previous research: when both women and men had German citizenship, they considered a religious funeral ceremony to be less important than those who held Turkish citizenship. Likewise, the higher the education of the respondents, the less likely they were to consider a religious funeral important.

The influence of the origin of the partner differed between women and men. The reference category is a partner who moved from Turkey to Germany as a first-generation migrant. Compared to male respondents whose partner migrated from Turkey, Turkish men attributed higher importance to a religious funeral when their partner was from Germany but had a Turkish or other migrant background. In contrast, if a Turkish woman in our sample lived with a partner from Germany who had a Turkish or other migrant background, she ascribed less importance to a religious funeral. For persons in unions where either partner was a German non-migrant and for respondents living alone, religious funerals were less important. These results suggest that partner choice impacts religiosity in a similar way to other indicators of socio-structural integration or

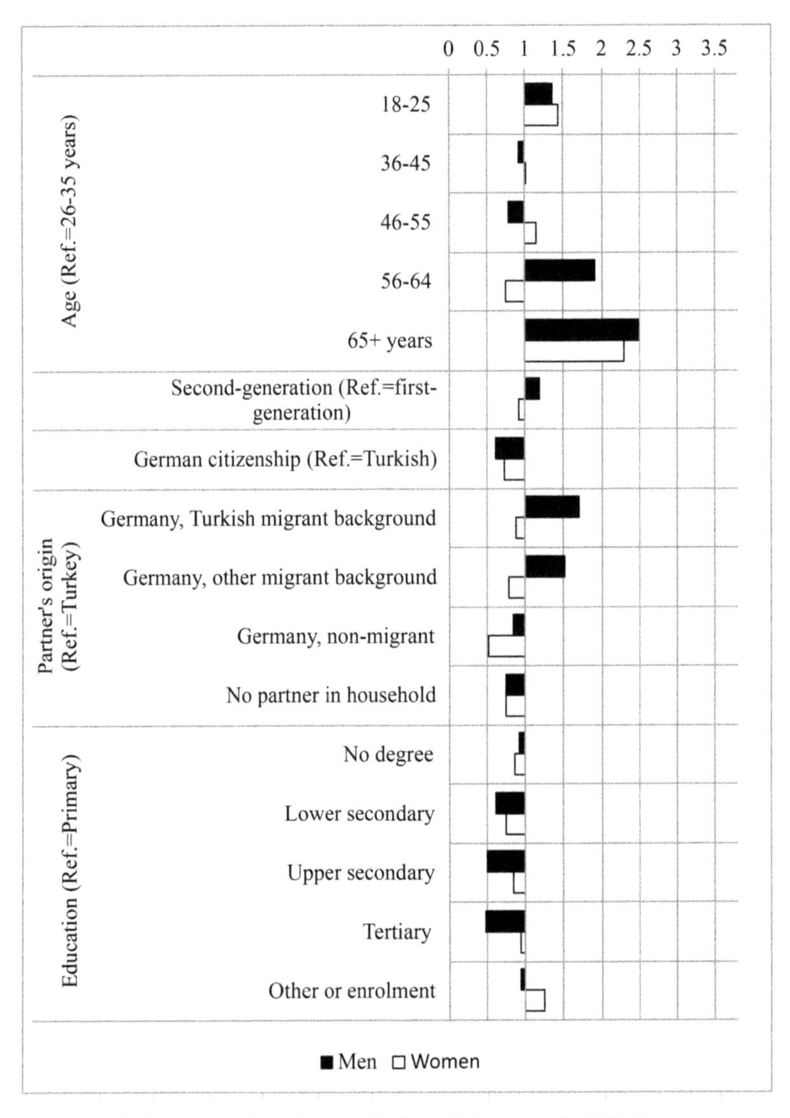

Source: Calculations based on German GGS/Turkish subsample (2005/06).
Note: N=4004; controlled for children living in household, missing values not displayed.

Figure 2. The importance of a funeral ritual among persons of Turkish migrant background in Germany, by sex (odds ratios).

secularisation, that is, single persons or those in mixed unions with a non-migrant/ migrant from a third country tend to be less religious than those living in unions where both have a Turkish background. This trend is even more pronounced among women than among men. It may well be that this is caused by a selection effect: less religious women may be more likely to have a partner from outside their origin group, or any partner, than more religious ones. For Turkish men, however, it makes a difference whether the partner is a migrant or not. Perhaps less religious men are more likely to live with a German partner than more religious men.

Associations between different religious indicators

As demonstrated, a religious funeral ceremony was the most important religious domain for Turkish women and men, at least for the questions posed. We then studied to what extent other religious indicators are associated with the importance of a religious funeral. Among those women and men who listed religion as one of their top-three child-rearing goals, 94 per cent also regarded a religious funeral ceremony important. About 97 per cent of those who found a religious marriage ritual important also attributed importance to a religious funeral ceremony. This is not too surprising, as a person who considers religion to be important at the beginning of the life course and at the time of family formation would also see religiosity playing a role at the end of life. Yet, many people who did not consider religion important in childrearing (84 per cent) or at marriage (54 per cent) said that a religious funeral ceremony is important. And no matter how often Turks attended places of worship, more than 90 per cent attached importance to a religious funeral. The share was only lower among those who said they never attended any religious meeting – but still 73 per cent ascribed importance to such a ceremony.

Conclusion

Our study employs representative survey data to analyse the impact of socio-demographic characteristics on attitudes towards the importance of a religious funeral ceremony among people of Turkish Muslim migrant background living in Germany. Although the dependent variable in our analysis is rather general, we can draw several conclusions: first, although religiosity may have declined and may continue to do so, Turkish migrants and their descendants place considerable weight on religiosity at the end of life: funerals are the most important religious tradition, with importance increasing the older people are. Second, although the socioeconomic integration of people of Turkish migrant background may increase with education and citizenship naturalisation, the importance attached to a Muslim funeral has remained stable from the first- to the second-generation. Third, this stability over generations can partly be explained by a transnational orientation, measured here as having a partner from Turkey who contributes to a 'renewal' of Turkish traditions. We should note, however, that funeral practices in Turkey are in flux and there are large regional and rural/urban differences (Türkis *et al.* 2012). This is to say that the attitudes of those moving from Turkey may not be more 'Turkish' than those of a person born to Turks in Germany.

Returning to the content of our dependent variable, the phrase 'a religious ceremony in a funeral' is rather broad. There is no way to know if survey participants were thinking of preparation of the deceased's body, different elements of the ceremony, characteristics of the gravesite, place of the funeral, or all of these items. Reading our results in combination with studies reviewed above we suggest the following interpretation: on one hand, religion and culture seem inseparable in life-course transition rituals such as funerals, therefore these will continue to be important for those in Turkish migrant communities who are religious. Among those with fewer or no religious ties, there may not be a distinction between religion and culture of origin, as ethnic/religious communities provide a familiar frame and infrastructure in an initially unknown context (Levitt 2003). Perhaps the work of the Turkish Funeral Funds provides not only economic incentives and support for

Turkish culture, but is regarded by the bereaved also as facilitating the logistical aspects of funerals.

On the other hand, we think that time will bring change. The social composition of the first-generation arriving today as well as subsequent generations born and raised in Germany is not comparable to that of the initial guest workers 50 years ago. The second- and subsequent generations intend to stay in Germany, meaning that the issue of where to be buried will certainly remain highly pertinent. We believe that change is likely to occur in this respect: despite the incentives for a funeral in Turkey, in the long run it may become more expensive to visit a parent's grave there than in Germany (in terms of money and of time spent travelling). Some members of the second- and third-generations have a non-migrant German parent, others a non-migrant German partner – they may therefore prefer to be buried here (Türkis *et al.* 2012, Balkan 2015a). Compromises between Muslim funerary doctrine and German legal conditions are also more readily available. The 250 cemeteries where Muslim funerals are possible may comprise only one per cent of all cemeteries in Germany, but given that most Turkish migrants are settled in the western part and primarily in cities (including Berlin), they may be in reach for most of these migrants today. Previous research on funerals (Jonker 1996, Ansari 2007) as well as on other life-course rituals, such as marriage (Aybek *et al.* 2015, Milewski and Huschek 2015), has demonstrated that Turkish migrants as well as other Muslim migrants in various destinations adjust their traditions to a transnational and a new societal context pragmatically without giving up their ethnic/cultural identity, with strong intergenerational ties and family values a central theme.

Jonker (1996) showed that there is variation by country of origin, although one cannot conclude the numbers of funerals for other Muslim groups from her study. We do know, however, that Turks are not the only Muslim immigrant group in Germany. To explore how the results of this and other studies on Turks can be extended to other groups, we carried out a few informal interviews with leaders and members of other Muslim communities in Germany, in which we discussed our findings. It seems that other immigrant groups do not maintain comparably high return intentions in general and therefore also do not consider post-mortem expatriation as frequently. This is particularly true for asylum seekers and refugees, but also for migrant workers and marriage migrants from other countries. One community leader[7] emphasised that local authorities are willing to find compromises – 'We found a pragmatic solution to use a coffin without actually using a coffin' – thus developing solutions which are acceptable to both sides.

This leaves us with a suggestion for further research. Work on funerary practices should focus not only on post-mortem mobility, but also on the adaptive processes at destination, their driving forces, conflicts, and solutions. This work should neither ignore heterogeneity among Muslims nor smaller and/or more recently-arrived migrant groups. We also suggest including not only the age of death in socio-demographic studies, but also data about religious funerary practices. Funerals reveal not only information about the individual who has passed away but are also telling for those still alive. Questionnaires could include items such as, 'Where are your parents buried?' or 'Where and how would you like to be buried?' The answers may tell us more about immigrants' identities and feelings of belonging than a single, straightforward question on national identity, and they may also shine a light on transitions and changes in increasingly multicultural and multi-religious societies.

Notes

1. In this paper, the term Turkey refers to the country of origin. The term migrants refers to the first-generation whereas their children are called descendants of migrants. Together they are referred to as having immigrant background, in accordance with official practice in Germany today (Swiazny and Milewski 2012). In official statistics, the term Turk(s) is based on citizenship. We do not differentiate further by ethnicity.
2. This survey is part of the Generations and Gender Program, which is coordinated by the United Nations Economic Commission for Europe (UNECE) in Geneva, and is designed to study demographic and social developments in 19 countries. In Germany, the GGS is carried out by the Federal Institute for Population Research (BiB), together with TNS Infratest (Ruckdeschel *et al.* 2006, Ette *et al.* 2007; see also: www.ggp-i.org).
3. Diehl and Koenig (2009) used the same data set as our study. They, however, created a summary indicator for the religious variables and combined the religious ceremonies at life events, that is, they did not explicitly focus on funerals.
4. As can be seen in the question on baptism, the GGS questionnaires were designed for primarily Christian contexts and used the same phrasing for the Turkish subsample. The question on the importance of a religious baptism ceremony is not applicable to a Muslim context, however. Although there may be 'welcome rites' for a newborn, a baptism ritual is generally not part of Muslim traditions (Dessing 2001, p. 25; McAuliffe *et al.* 2001). Hence, this question was not used in our study.
5. Odds ratios lower than 1 indicate that the respective group is less likely than the reference group to think funerals important; values above 1 indicate that the respective group attributes higher importance to religious funerals than the reference group.
6. About 61 per cent of the respondents lived in a household with their children. This variable did not have a significant statistical impact (not displayed).
7. Informal interview with the leader of a local community centre in Eastern Germany in December 2015. The members originate from various Arabic countries.

Acknowledgments

We thank two anonymous reviewers and Eva Soom Ammann and Alistair Hunter, the guest editors of this special issue, for their constructive feedback and helpful suggestions as well as Renée Lüskow for language editing.

Disclosure statement

No potential conflict of interest was reported by the author(s).

Funding

This study was supported by the first author's European Reintegration Grant provided by Marie Curie Actions (FP7 People, PERG-GA-2009–249266 – MigFam) and funded by the European Commission. The views expressed in this paper do not reflect the views of these funding agencies.

References

Abadan-Unat, N., 1995. Turkish migration to Europe. *In*: R. Cohen, ed. *The Cambridge survey of world migration*. Cambridge: Cambridge University Press, 279–284.

Alba, R.D., and Nee, V., 2003. *Remaking the American mainstream: assimilation and contemporary immigration*. Cambridge: Harvard University Press.

Ansari, H., 2007. Burying the dead: making Muslim space in Britain. *Historical research*, 80 (210), 545–566.

Aybek, C., Straßburger, G., and Yüksel-Kaptanoglu, I., 2015. Marriage migration from Turkey to Germany: risks and coping strategies of transnational couples. *In*: C. Aybek, J. Huinink, and R. Muttarak, eds. *Spatial mobility, migration, and family dynamics*. Dordrecht: Springer, 23–42.

Balkan, O., 2015a. Burial and belonging. *Studies in ethnicity and nationalism*, 15 (1), 120–133.

Balkan, O., 2015b. Till death do us DEPART: repatriation, burial, and the necropolitical work of Turkish funeral funds in Germany. *In*: Y. Suleiman, ed. *Muslims in the UK and Europe*. Cambridge: Cambridge University Press, 19–28.

Baykara-Krumme, H., 2013. Intergenerational relationships in old age: Turkish families in Turkey and in Western Europe. *Journal of family research*, 25 (1), 9–28.

Berger, P., and Luckmann, T., 1979. *The social construction of reality*. London: Penguin.

Bundesamt für Migration und Flüchtlinge (BAMF), ed., 2011. *Migrationsbericht 2010*. Nürnberg: BAMF.

Campo, J.E., 2006. Muslim ways of death. Between the prescribed and performed. *In*: K. Garces-Foley, ed. *Death and religion in a changing world*. New York: M.E. Sharpe, Inc., 147–177.

Dessing, N.M., 2001. *Rituals of birth, circumcision, marriage, and death among Muslims in the Netherlands*. Leuven: Peeters.

Destatis – Statistisches Bundesamt/Federal Statistical Office. 2012. *Bevölkerung und Erwerbstätigkeit. Natürliche Bevölkerungsbewegung. Fachserie 1. Reihe 1.1*. Wiesbaden: Destatis.

Destatis – Statistisches Bundesamt/Federal Statistical Office. 2014. Zensusdatenbank. https://ergebnisse.zensus2011.de/ [Accessed 23 May 2015].

Diehl, C., and Koenig, M., 2009. Religiosity of first and second generation Turkish migrants. A phenomenon and some attempts at a theoretical explanation. *Zeitschrift für Soziologie*, 38 (4), 300–319.

El-Menouar, Y., 2014. The five dimensions of Muslim religiosity. Results of an empirical study. *Methods, data, analyses*, 8 (1), 53–78.

Ette, A., *et al.*, 2007. *Generation and gender survey. Dokumentation der Befragung von türkischen Migranten in Deutschland*. Wiesbaden: BiB.

Foner, N., and Alba, R., 2008. Immigrant religion in the U.S. and Western Europe: bridge or barrier to inclusion? *International migration review*, 42 (2), 360–392.

Gans, H.J., 1979. Symbolic ethnicity: the future of ethnic groups and cultures in America. *Ethnic and racial studies*, 2, 1–20.

Hamel, C., *et al.*, 2012. Union formation and partner choice. *In*: M. Crul, J. Schneider and F. Lelie, eds. *The European second generation compared: Does the integration context matter?* Amsterdam, Chicago: Amsterdam University Press, 225–284.

Haug, S., Müssig, S., and Stichs, A., 2009. *Muslim life in Germany*. A study conducted on behalf of the German Conference on Islam. Nürnberg: Federal Office for Migration and Refugees, Research Report 6.

Höpp, G., and Jonker, G., eds., 1996. *In fremder Erde. Zur Geschichte und Gegenwart der islamischen Bestattung in Deutschland*. Berlin: Verlag Das Arabische Buch.

Jonker, G., 1996. The knife's edge: Muslim burial in the diaspora. *Mortality: promoting the interdisciplinary study of death and dying*, 1 (1), 27–43.

Kaplan, O., 2004. *Tod im Islam: Und Sterben in der Türkei*. Frankfurt: Y. Landeck.

Karakasoglu, Y., 1996. Bestattung und Türkisch-islamische Organisationen. *In*: G. Höpp and G. Jonker, eds. *In fremder Erde. Zur Geschichte und Gegenwart des islamischen Bestattung in Deutschland*. Berlin: Verlag Das Arabische Buch, 83–105.

Krumme, H., 2004. Fortwährende Remigration: Das transnationale Pendeln türkischer-Arbeitsmigrantinnen und Arbeitsmigranten im Ruhestand. *Zeitschrift für Soziologie*, 33, 138–153.

Kuhnen, C., 2009. *Fremder Tod. Zur Ausgestaltung und Institutionalisierung muslimischer, jüdischer, buddhistischer, hinduistischer und yezidischer Bestattungsrituale in Deutschland unter dem Aspekt institutioneller Problemlagen und gesellschaftlicher Integration*. Universität Bremen (Diss.).

Larose, D.T., 2006. *Data mining methods and models*. Hoboken, NJ: J. Wiley & Sons.

Levitt, P., 2003. 'You kow, Abraham was really the first immigrant': religion and transnational migration. *International migration review*, 37 (3), 847–873.

Maliepaard, M. Lubbers, M., and Gijsberts, M., 2010. Generational differences in ethnic and religious attachment and their interrelation. A study among Muslim minorities in the Netherlands. *Ethnic and racial studies*, 33 (3), 451–472.

McAuliffe, J.D. *et al.*, 2001. *Encyclopedia of the Qu'rān. Volume One A-D*. Leiden: Brill.

McAuliffe, J.D., *et al.*, 2002. *Encyclopedia of the Qu'rān. Volume Two, E-I*. Leiden: Brill.

Metcalf, P., and Huntington, R., 1991. *Celebrations of death. The anthropology of mortuary ritual*. 2nd ed. New York: Cambridge University Press.

Milewski, N., and Huschek, D., 2015. Union formation of Turkish migrant descendants in Western Europe: family involvement in meeting a partner and marrying. *In*: N. Milewski, I. Sirkeci, M.M. Yücesahin and A. Rolls, eds. *Family and human capital in Turkish migration*. London: Transnational Press, 11–23.

Phalet, K., and Schönpflug, U., 2001. Intergenerational transmission of collectivism and achievement values in two acculturation contexts: the case of Turkish families in Germany and Turkish and Moroccan families in the Netherlands. *Journal of cross-cultural psychology*, 32 (2), 186–201.

Phalet, K., Fleischmann, F., and Stojcic, S., 2012. Ways of 'being Muslim': religious identities of second-generation Turk. *In*: M. Crul, J. Schneider and F. Lelie, eds. *The European second generation compared: does the integration context matter?* Amsterdam: Amsterdam University Press, 341–374.

Phalet, K., Gijsberts, M., and Hagendoorn, L., 2008. Migration and religion: testing the secularisation thesis among Turkish and Moroccan Muslims in the Netherlands 1998–2005. *In*: F. Kalter, ed. *Migration, Integration und Ethnische Grenzziehungen. Kölner Zeitschrift für Soziologie und Sozialpsychologie*, Sonderheft 48. Wiesbaden: VS, 412–436.

Pries, L., 2010. *Transnationalisierung. Theorie und Empirie grenzüberschreitender Vergesellschaftung*. Wiesbaden: VS.

Reimers, E., 1999. Death and identity: graves and funerals as cultural communication. *Mortality*, 4 (2), 147–166.

Ruckdeschel, K., *et al.*, 2006. Generations and Gender Survey. Dokumentation der erstenWelle der Hauptbefragung in Deutschland. *Materialien zur Bevölkerungswissenschaft*, 121a. Wiesbaden: BiB.

Rumbaut, R.G., 2004. Ages, life stages, and generational Cohorts: decomposing the immigrant first and second generations in the United States. *International migration review*, 38 (3), 1160–1205.

Smits, F., Ruiter, S., and van Tubergen, F., 2010. Religious practices among Islamic immigrants: Moroccan and Turkish men in Belgium. *Journal for the scientific study of religion*, 49 (2), 247–263.

Strumpen, S., 2012. Altern in fortwährender Migration bei älteren Türkeistämmigen. *In*: H. Baykara-Krumme, A. Motel-Klingebiel and P. Schimany, eds. *VieleWelten des Alterns. Ältere Migranten im alternden Deutschland*. Wiesbaden: Springer VS, 411–433.

Swiazny, F., and Milewski, N., 2012. Internationalisierung der Migration und demographischer Wandel. EineEinführung. *In*: B. Köppen, P. Gans, N. Milewski and F. Swiazny, eds. *Internationalisierung: Die unterschätzte Komponente des demographischen Wandels in Deutschland*. Schriftenreihe der DGD, 5. Norderstedt: BoD, 11–41.

Tan, D., 1998. *Das fremde Sterben: Sterben, Tod und Trauer unter Migrationsbedingungen*. Frankfurt: IKO.

Türkis, I., and Seeberger, B., 2012. Islamisches Bestattungsverhalten und Trauerrituale in der Migration - aufgezeigt an türkischen Migranten. *In*: F. Karl, ed. *Transnational und translational. Aktuelle Themen der Alternswissenschaften*. Berlin: LIT, 57–75.

Türkis, I., Meiners, N., and Seeberger, B., 2012. Islamische Trauerrituale in der Fremde. Eine Untersuchung, aufgezeigt an türkischen Migranten in Deutschland. *HeilberufeSCIENCE*, 3, 119–125.

Van Tubergen, F., 2006. Religious affiliation and attendance among immigrants in a secular society. A study of immigrants in the Netherlands. *Journal of ethnic and migration studies*, 33 (5), 747–765.

Van Tubergen, F., and Sindradóttir, J.I., 2011. The religiosity of immigrants in Europe: a cross-national study. *Journal for the scientific study of religion*, 50 (2), 272–288.

Zirh, B.C., 2012. Following the dead beyond the 'nation': a map for transnational Alevi funerary routes from Europe to Turkey. *Ethnic and racial studies*, 35 (10), 1758–1774.

Staking a Claim to Land, Faith and Family: Burial Location Preferences of Middle Eastern Christian Migrants

Alistair Hunter

Islamic and Middle Eastern Studies, University of Edinburgh, Edinburgh, UK

ABSTRACT

The question of where to conduct funeral rituals may confront migrants and their descendants with a stark existential choice which reveals much about how identities are negotiated in and through place. This paper scrutinises the relationship between identity and place through the prism of preferred burial location. More concretely, it sets out a typology of motivations for preferred burial location in contexts of migration. In addition to advancing analytical clarity with this typology, the paper also aims to promote theoretical clarity by questioning the hypothesis that burial in the country of residence constitutes a straightforward indicator of migrant integration. Based on 67 qualitative interviews with Christians of Middle Eastern origin in Britain, Denmark and Sweden, the paper presents various rationales for preferred burial location, showing the sometimes ambivalent relationship which migrants negotiate between place and identity.

The study of death and dying in migration contexts is a relatively recent development in European scholarship on migration and ethnic minorities, with a small number of scholars first engaging in this field around the turn of the twenty-first century (Jonker 1996, Tan 1998, Reimers 1999, Chaïb 2000, Gardner 2002). One question which this body of literature has treated, albeit partially and disparately, is the preferred place of burial (or other form of disposal). In other words, whether to choose the country of origin, the country of residence, a third country, or indeed the transnational option of conducting funerary rituals in more than one location. Although hitherto overlooked by scholars, both in death studies and migration studies, I argue in this paper that burial choices in the context of migration can reveal much about the connection between place and identity. My aim here is first to stimulate reflection on the topic of burial choice in migration contexts, in anticipation of the wider set of questions flowing from transnational ageing and dying that such reflection may elucidate: the care of vulnerable strangers, the ethics of hospitality and the 'eschatological questions that speak to us all. "Who am I?", "How did I get here?"' (Gunaratnam 2013, pp. xiv–xv). These questions will only become more salient in the decades to come, according to the latest demographic projections which show substantial increases in the number of older migrant populations in the coming decades (Rallu 2016). More concretely, this paper contributes to this promising line of study by presenting a systematic typology of motivations for preferred burial location. Such a typology will serve, I hope, to bring greater analytical clarity to scholarly analyses of the relationship between place and identity through migrant funerary rituals.

In conceptual and theoretical terms, the connections between place and identity have been most thoroughly explored by geographers and environmental psychologists. In the latter discipline, Proshansky *et al.* (1983) developed a theory of place–identity which argued for the centrality of environment-related awareness in the development of self-identity. The geographer Cresswell has defined place as 'meaningful space' (2004). Conceiving of place as meaningful space underlines the direct connection between place and identity, particularly in contexts of post-migration diversity and inter-ethnic relations, as another geographer John Clayton observes: 'Identities do not just take place, but also make place' (Clayton 2009, p. 483). The increased capacity of migrants from diverse socio-economic backgrounds to maintain multiple ties with 'home' places and with other nodes in their transnational networks, thanks to the democratisation of long-distance travel and communication technologies, means that questions of migrant place-making are ever more complex and fluid (Kaplan and Chacko 2015).

Just as the act of migrating from one place to another may constitute a major turning point in the lifecourse, death signifies an even more radical, indeed the most radical, juncture, bringing into sharp focus the relationship between place and identity. Reimers (1999) reminds us that in many cultures ancestral burial grounds are a privileged place, synonymous with 'home' or place or origin. The anthropologist Myrna Tonkinson shows how funerals have gained in significance for Indigenous Australian communities in recent decades. In the face of mounting inequalities (including in terms of life expectancy), Indigenous Australian funerals – often attended by hundreds of mourners – have become 'settings for the display of solidarity, the assertion of identity and autonomy, and the expression of a determination to retain their distinctiveness' (2008, p. 52). In social and cultural geography, a literature on 'deathscapes' has emerged, defined as 'the material expression in the landscape of practices relating to death' (Teather 2001, p. 185). Deathscapes are not only the terrain of the dead and dying, but are also intensely meaningful – if contested – places where the living find a 'spatial fix' for mourning and memorialisation (Hallam and Hockey 2001, Maddrell and Sidaway 2010).

Given the rich ground for reflecting on the relations between place and identity which the contexts of migration and death respectively provide, it is noteworthy that few scholars have sought to combine the insights from these two bodies of literature. The deathscapes literature can be critiqued for its lack of engagement with the diversity of funerary practices resulting from international migration (Hunter 2015). Likewise the literature on migrant place-making has rarely engaged with sites of funeral practices or memorialisation. Geographic mobility leads people to identify with multiple places, and consequentially opens up multiple options for where people envisage their final resting place to be (Casal *et al.* 2010). Death in migration is an interesting point at which to study these questions of place and identity. Particularly for cultures which dispose of the dead through burial – the focus of the present analysis – the choice of location may provoke profound questions of identity: 'In the choice of place of burial, the soil/earth becomes a fundamental "where", a stable basis by which the place of origin is precisely defined' (Chaïb 2000, p.

24; author's translation). As I argue elsewhere (Hunter 2015), death is a critical juncture in the settlement process of migrant families and communities. In terms of location of memorialisation and disposal rites, the individual or the bereaved have three main choices open to them.

First, death can be an occasion to emphasise self-conceptions of temporary presence and 'guesthood', by opting for posthumous repatriation and conducting funeral rites in countries of origin. While alive, migrant elders may nurture ambivalent feelings about the legitimacy of their place both in countries of origin and immigration. However, as Katy Gardner recounts in her study of Bengali seniors in London, approaching death there may occur:

> a significant shift in attitude. As the domain of the sacred and – inextricably – the site where the patrilineage is based […] it is to [the homeland] that most elders feel their corpses, if not always their living bodies, should return. (2002, p. 205)

It is no surprise therefore that in many countries of immigration a veritable industry has grown up to facilitate repatriation of deceased migrants (Chaïb 2000). In contrast, for some it becomes harder to 'stake a territorial claim' via burial in countries of immigration (Ansari 2007, p. 563), particularly due to apprehensions about non-observance of religious funerary rituals (Chaïb 2000, Venhorst 2013).

Second, death can be an occasion to lay what are perhaps the deepest foundations for settlement and belonging, through ritual practices in the country of immigration. In Reimers' study, the graves of Serbs in Sweden became a mooring for the community there (1999). Others go one step further to argue that the act of burial in the country of residence should be interpreted as the ultimate (in all senses of the word) marker of migrant integration (Chaïb 2000, Oliver 2004); or to put it differently, in Chaïb's neat turn of phrase, 'the integration of […] immigrant communit[ies] through the disintegration of their corpses' (2000, p. 29; author's translation).

Third, migrants and their descendants may also opt to conduct funeral rites both in places of origin and settlement. This can be interpreted as a quintessentially transnational solution to the questions of place and identity posed above. Yet the literature on transnational funerary rituals is sparse and primarily limited to studies of groups which practice cremation due to the greater portability of cremated remains (Jassal 2015).

Pioneering quantitative research has been conducted in France showing the varied significance of each of these three choices. Based on a survey of over 6000 older migrants in France, Attias-Donfut and Wolff (2005) propose three main sets of factors which influence preferred burial location: territorial attachments to 'home' and 'host' countries; religious affiliation; and family attachment. Although not encompassing international migrants, evidence from Spain and France (Casal *et al.* 2010) and Sweden (Marjavaara 2012) also shows the importance of family, religious and territorial attachments in decisions about place of burial. Marjavaara's study makes a rare contribution insofar as post-mortal mobility decisions are analysed from data on actually accomplished burials rather than the stated preferences of living respondents to questions about future burial location: as with other forms of bodily mobility, there is always likely to be a discrepancy between stated preferences and actual outcomes (see also Rowles and Comeaux 1987). Data on actual outcomes has a particularly useful practical application in estimating future demand for burial space in different locales. By contrast the advantage of qualitative studies based on prospective preferences, such as this one, lies instead in unpicking the nuances of the relationship between place and identity as experienced by the living. In what follows I will present a typology of motivations for preferred burial location, building on the works cited above and based on an analysis of 67 interviews with Middle Eastern Christians resident in the UK, Denmark and Sweden. In the Discussion, I will return to the question of integration through burial, critically evaluating the contentions made in the literature against the data collected for this paper. Before that, however, some background information about the methodology and populations under study will provide the necessary context to the research.

Research context and methodology

Burial is not a disposal method which is universally practised: the generalisability of the present analysis is therefore limited to those migrant communities in which burial is the norm. Burial is understood here as including both whole body inhumation and the interment of remains such as ashes. Reviewing the small number of studies which have included the question of preferred burial location, it transpires that Muslim migrant communities have been the primary focus of attention (Jonker 1996, Chaïb 2000, Gardner 2002, Ansari 2007, Venhorst 2013). In adding a new empirical dimension to this literature, I focus here on the burial preferences of Middle Eastern Christian migrants, comparing Egyptian (Coptic Orthodox), Assyrian and Iraqi Christians (various denominations) in three European countries: Denmark, Sweden and the UK.

Within these ancient Middle Eastern churches, eschatological questions about the condition of the soul and the body after death, and the chronology of future salvation, were settled as early as the fourth century. As such, 'doctrine on the last things differs little' among these denominations: whole body inhumation is recommended in preference to other means of disposal (such as cremation) owing to belief in bodily resurrection at the time of the Last Judgment (Cody 1991, pp. 973a–974b). Prayers of intercession for the dead are a common liturgical feature, and visiting graves to pray for deceased relatives takes place around important dates in the church calendar such as Christmas Eve, Easter and Ascension Day (Wissa-Wassef 1991).

In Europe, Middle Eastern Christian communities have developed from a diverse range of migration routes including postcolonial and guestworker flows, high skilled labour and student migration, as well as significant refugee movements. This study focuses on Middle Eastern Christian communities in Denmark, Sweden and the UK. These countries were chosen in the context of the wider project *Defining and Identifying Middle Eastern Christians Communities in Europe* (DIMECCE), on the basis of their analytical value in comparing variations in church-state models and different community sizes and patterns of settlement. In the UK, the Coptic Orthodox Church is by far the largest Middle Eastern Christian community with over 20,000 adherents. Iraqi and Assyrian Christians number in total some 8–10,000 people in Britain. Iraqi and Assyrian groups are less geographically spread than the Copts, with a clustering of communities, churches and other institutions in West London. The situation is different in Sweden where a more open refugee policy has led to a significant number of Middle Eastern Christians settling in the country. The vast majority – some 120,000 people – are Assyrians/Syriacs (*Assyrier/Syrianer*) originally from Turkey, Lebanon, Iraq and Syria, and most belong to the

Syriac Orthodox Church. In addition, there are a few thousand Copts and some 20,000 members of the Chaldean Catholic Church, most of whom fled Iraq due to the security situation in the 1990s and since 2003. For all these communities, the small city of Södertälje near Stockholm is a particular hub. In Denmark, the community context differs again with the majority of Middle Eastern Christians being of Iraqi origin. In 2014, there were around 3000–3500 Christians of Iraqi background, and 500–600 Christians of Egyptian origin. Most live in the Copenhagen and Århus areas. Among the Christians from Iraq, many are Assyrians or Chaldeans who fled the wars in Iraq from the 1980s onwards. Most of the Egyptian Copts arrived as migrant workers between the late 1960s and 1980s. Due to the small numbers in Denmark, not all communities have access to their own church or even their own priest, something which impacts on their practices and identifications (Galal *et al.* 2016).

The data on which this analysis is based derives from 67 semi-structured in-depth interviews with members of the above communities. These interviews were conducted for the DIMECCE research project by myself and colleagues at the universities of Łodz, Roskilde and St Andrews, between February and July 2014. The DIMECCE project's focus on 'defining' Middle Eastern Christian identities in Europe dictated our sampling frame: we selected our interviewees from among those Middle Eastern Christians who were identifiable as having an active role in defining and/or representing their community or congregation. These figures included clergy (bishops and priests), deacons, lay representatives (e.g. members of church boards), Sunday school teachers, church youth leaders, political activists, and representatives of cultural, charitable and sports associations. Doctrinal injunctions against women being given ordained roles in their churches limited the number of females in active representative roles whom we could interview, hence the ratio of 49 men to 18 women in the sample.

Respondents were asked about burial location preferences in the interview's final section concerning connections to 'the homeland' – a necessarily vague term for some respondents given their forebears' experiences of redrawn borders and displacement within the Middle East. Unless the interviewee had volunteered the information previously, the question about burial location preferences was generally asked following a series of questions on the theme of homeland return. Asking about burial preferences and the possibility of posthumous return – a potentially sensitive topic – came more naturally at this juncture, by which time also a sufficient degree of rapport had been established between the interview participants.

The sample was split between first- (including 1.5 generation) and second-generation respondents. Forty-nine first-generation migrants were asked where they would prefer to be buried when they come to the end of their lives, in the place of origin or country of residence. Following their initial response, they were then prompted for the motivations behind their answers. In addition, 18 second-generation descendants were asked whether the question of burial location was a topic which is discussed in their families by older relatives; in addition to giving answers on these family dynamics, some second-generation respondents also volunteered their own preferences regarding burial location.

The interviews were transcribed and the resulting transcripts were imported into the NVivo software package for qualitative data analysis. Analysis was an iterative process, informed by the insights of previous work, in particular the distinction between religious, territorial and familial motivations for preferred burial location (Attias-Donfut and Wolff

2005, Casal *et al.* 2010, Marjavaara 2012). However, repeated reading of the transcripts revealed further nuances within these categories, as will be presented below.

A typology of motivations for preferred burial location in migratory contexts

Table 1 presents a typology of motivations for preferred burial location in migratory contexts. In what follows I will elaborate each type of motivation, supported by examples from interview data. A first point to note however, is that a small minority of respondents were completely indifferent to the location of their final resting place. To quote one respondent, 'What's the difference? We're all – it's all one ground. If you're gonna decay, you're gonna decay my friend, it doesn't matter where you are [laughs]' (Assyrian/Syriac, Male, 30s, 1.5-generation, UK). Such nonchalance may be read as genuine indifference or alternatively as a coping mechanism to deflect ontological insecurities around mortality.

Supplementing the three sets of variables identified in existing literature – familial, territorial and religious – I begin here with a prior set of practical considerations which emerged in interviewees' narratives of preferred burial location, namely financial and organisational costs, and appraisals of the security situation in a given location. This has been mentioned briefly as a factor influencing burial location (Attias-Donfut and Wolff 2005, Venhorst 2013), but in this analysis it enters centre stage.

Table 1. Typology of motivations for preferred burial location.

Practical considerations	Family considerations	Territorial considerations	Sacred considerations
Financial costs	Ancestors	Preponderant presence	Theological conformity
Organisational costs	Parents	The nation	Religious disassociation
Security situation	Nearest and dearest	Emotional landscape	Sacralising
	Descendants		

Practical considerations: striking a balance between money, coordination and safety

Financial costs expresses a preference based on pragmatic consideration of the economic outlays associated with burial in the countries of origin and immigration, respectively. The significant transport costs of repatriation to the Middle East, where cemetery plots nonetheless remain free or easily affordable, may effectively be offset by the high costs associated with burial in many European countries, especially the purchase of burial plots in perpetuity but also including expenditure for coffins, hearses, catering and receptions for mourners. Indeed, the elevated costs of burial in London had recently prompted representatives of the Assyrian community there to institute a funeral cooperative fund. According to one of the founders of this initiative:

> We collect membership from people, very little money – membership – and then we spend up to £4000 for every funeral. We pay for the coffin, we pay for the plot, we pay for limousines, hearse, all that. And then the person – the people [who] lose their father or brother or sister or whatever – in the house, they don't have to do nothing. (Assyrian/Syriac, Male, 50s, first-generation, UK).

The last point raised here, that the recently bereaved can grieve in peace in the family home without having to attend to the funeral logistics, brings to light the *organisational costs* involved in arranging a funeral. Organisational costs have rarely been discussed as a factor influencing preferred burial location in the existing literature:[1] nonetheless, they appear as decisive for some respondents. As a UK-resident Copt noted, 'Not to make a hassle on my family, here is much better' (Coptic, Male, 40s, first-generation, UK). On the one hand, repatriation is not only costly in financial terms but involves a lot of administrative form-filling, as well as extended periods of bereavement leave for those who accompany the casket. On the other hand, bureaucratic requirements may be minimal once the deceased has arrived in the place or origin: 'You know, London, it's not like Iraq, just paperwork from the church and then they say "go bury him". Here, there is the cemetery's involvement, the funeral director's involvement, council tax involvement, [Municipal] Council, town hall, the hospital's involvement' (Assyrian/Syriac, Male, 50s, first-generation, UK).

Security situation – the third practicality which emerged in interviews – resonated with respondents from specific national backgrounds. In particular, the fragile security situation for Christians in Iraq at the time of the interviews was mentioned by several respondents:[2]

> Most of them don't see that they would move back [to Turkey or Iraq] ... They don't because it's not safe to take them there. (Assyrian, Female, 40s, first-generation, Sweden)

> You can't go back, you can't practically go back to Iraq no. (Iraqi, Male, 50s, first-generation, UK)

However, some Iraqi respondents refused to be intimidated by the worsening security situation in their homeland. The following quote shows the primary importance of family ties over security concerns when it comes to final resting place: 'So I want to be repatriated. I also told my children [...] I want to travel over there, even though ISIS (see note 2) is there because I have family – I have my brother's grave, my father's [grave]' (Iraqi, Female, 50s, first-generation, Denmark). In the next section I will present the family-based rationales for preferring one location over another.

Family considerations: genealogical continuity or the 'new first ancestors'

From time immemorial, cemeteries have served the function of regrouping successive generations of a kin group around a common ancestor. This practice has been interpreted as a means of maintaining ontological security by reconciling the individual with the finite nature of existence (Reimers 1999). The importance of family and kin ties in burial thus serves a double purpose: looking backwards in time by assuring the continuity of the group through genealogical affiliation (Chaïb 2000); and looking to future generations to keep alive the memory of forebears who have died, that is, survival by proxy (Casal *et al.* 2010). The act of emigrating from the place of origin, which is the location of the ancestors, inevitably disrupts this mechanism. First-generation migrants either have to opt for repatriation and burial among the ancestors, but at the risk of being lost to the 'memory work' of future generations; or to break with the genealogy in the place of origin in the hope of becoming the 'new first ancestor' for future generations in the adopted homeland.

Broadly, four family-based considerations corresponding with different temporal orientations can be discerned: an orientation to the past, via (i) ancestors in the place of origin, or (ii) via deceased parents, laid to rest either in the ancestral soil or the adopted homeland; (iii) an orientation to 'significant others' in the present or (iv) an orientation to future generations in the country of settlement. The *ancestors* category in the typology above expresses a preference for burial in the country of origin in order to maintain affiliation with the genealogical line going back in time. It may be associated with the existence of a family burial ground or family mausoleum:

> In Egypt, in my family's mausoleum. Because it's such an ancient land, and I belong to it. It's a very, very romanticised, irrational view, but yeah, that's the reason why. (Coptic, Female, 40s, 1.5-generation, UK).

The category of *parents* is also oriented to the past, more specifically the recent past, namely a desire to be buried next to where deceased parents have been laid to rest (see also Attias-Donfut and Wolff 2005). However, unlike burial location choices based on ancestral orientation, a parental orientation may favour burial either in the country of origin – as in the quote above from the woman who disregarded the dire security situation in Iraq – or in the country of settlement:

> There is no point [that my children] should take me back home or bury me there. And that will not only be a [financial] cost, actually – because my father and my mother are buried here, as well, and so … I don't want them to bother really. (Assyrian/Syriac, Male, 70s, 1.5-generation, UK)

Also very salient were the preferences expressed either for burial beside a deceased spouse, sibling(s), child(ren), close friends and so on, or in being laid to rest where a living spouse, sibling(s), child(ren), and close friends can easily visit the grave and offer their prayers. This set of considerations is termed *nearest and dearest* in my analysis. Temporally speaking it characterises a perspective oriented to the present, to those 'significant others' who are part of respondents' daily lives: 'I don't have any special wishes to be buried in Iraq … the most important for me, is that my kids or relatives should have the possibility to visit me' (Iraqi, Male, 30s, first-generation, Denmark).

A more radically future-oriented conception of memorialisation is seen in responses expressing a preference based on affiliation with the genealogical line, but projecting forward in time towards future *descendants* rather than back in time towards ancestors. For the first-generation, such a burial location preference indicates a respondent's self-understanding as 'the new first ancestor' in the country of immigration; second-generation respondents can also express an orientation to descendants. In the following example, a young Iraqi male, who grew up in Denmark, credits his grandmother as the 'new first ancestor' whom other relatives have 'gathered around':

> […] well I have my grandmother who – who lived in Germany, but because she has 3 daughters here in Denmark she actually chose to be buried in Denmark, so she is buried in [name of] churchyard. There where one of my aunts lives, so then it is like gathering around it. (Iraqi, Male, 20s, second-generation, Denmark)

I have shown that when discussing the familial considerations influencing preferred burial location, temporal perspectives are paramount. The question of time is also crucial when discussing territorial attachments in terms of burial: the weight of the years spent in a given place has a heavy influence.

Territorial considerations: at peace with place

The category of *preponderant presence* expresses significant attachment to territory defined either in individual terms as the time spent in a given location or in collective terms as the size of the community which is present in that given location. The term preponderant is chosen to indicate not just a quantitative significance but also a qualitative significance, for example, for young people, spending their formative years in a given location. Collectively, the geographical weight of numbers of a community, be that a community of the living or the dead, can be a strong enticement to consider burial in that location.

It was mentioned earlier that Södertälje, in Sweden, is a major population centre for Assyrian/Syriacs as well as other Middle Eastern Christian denominations in Scandinavia. Given the high proportion of Södertälje's population who are Assyrian/Syriac, burial there becomes more self-evident, as recounted by a young Assyrian women: 'I think that there are so many people here that [in death] they will feel home here too' (Assyrian/Syriac, Female, 20s, 1.5-generation, Sweden). In individual terms, the significance of a given location for burial was often recounted in terms of the time spent and invested by individuals there:

> In the UK, not in Egypt, no. Because I've lived here all my life. (Coptic, Female, 40s first-generation, UK)

> Bury me here, by all means. For no reason whatsoever but just being casual and thinking quick off my head given I've had two thirds of my life here and one third there. (Iraqi, Male, 60s, first-generation, UK)

Narratives invoking *the nation*, broadly defined, featured in many respondents' rationales for choosing one national soil over another. In explaining his preference to be buried in Britain, a dual national Coptic respondent contrasted the discomfort he feels regarding his Egyptian nationality and the ease with which he assumes a British belonging:

> I'm British. In fact I'm ashamed to say that my Egyptian passport has expired and I haven't bothered to renew it. And the reason is if I have to use it, I use it once a year for three days, four days I spend in Egypt but I use my British passport everywhere, I'm very comfortable with it and eh I am very loyal to Britain ... Anyone ask me a question at the airport, for example, I'm British. (Coptic, Male, 50s, first-generation, UK)

Similarly, a female, Assyrian/Syriac respondent eloquently expressed her attachment to Denmark by highlighting the tranquillity which she had found there and to which she aspired for her final resting place:

> Here in Denmark, 100 percent yes, I'm not even in doubt about it, because you can believe me when I say, that – I feel this is my country, because I have received the things I was lacking. It is very important what I am saying, it is peace and quiet, what you wish for. (Assyrian/Syriac, Female, 50s, first-generation, Denmark)

Other respondents expressed a burial location preference based not on national characteristics but rather on more emotional or aesthetic resonances with (or rejections of) a specific place or landscape. In the above typology I use the term *emotional landscape* to describe this type of narrative. In other studies, emotional landscapes may correspond with where individuals have bought second homes (Marjavaara 2012). Some 1.5- and second-generation individuals mentioned happy childhood holiday memories visiting historic or scenic tourist sites in their parents' places of origin. Another simply stated: 'we have a lovely cemetery here' (Coptic, Male, 60s first-generation, UK). Conceiving of cemeteries as places of tranquillity and transcendent beauty marks an appropriate introduction to the final part of this section, in which I explore the specifically sacred dynamics of place-making through diasporic burial.

Sacred considerations: pious indifference or purposeful inscription of space

Describing the funeral rituals performed by her Coptic congregation in Sweden, a young first-generation female, said:

> [The priest] of course has to say the prayer over the deceased, "you don't miss that person, he's now in a good place" and those things. And it's the same in Egypt … here it's like in Egypt. We are trying … we are trying to keep the tradition … we are trying to do the same … . (Coptic, Female, 20s, first-generation, Sweden)

This effort to 'do the same' as in the homeland mother church is an orientation which I label *religious conformity*: it expresses a preferred burial location based on conformity with formal and codified religious ritual practices. Equally, as Marjavaara (2012) notes, burial location preferences may be a reaction against religious affiliation in circumstances where the deceased or bereaved are negatively disposed to the religious institutions in which they were brought up. I label this as *religious disassociation*. Perhaps unsurprisingly, given the focus of the DIMECCE project on individuals who have an active role in representing their religious communities, this motive did not emerge in interviews. It is nonetheless included here so that the typology may better integrate existing insights from the literature.

Although the quantitative analysis by Attias-Donfut and Wolff (2005) shows that personal religiosity is correlated with a preference for burial in the country of origin, qualitative work has shown that it may equally induce a preference for burial in the country of immigration (Ansari 2007). In this regard much depends on the regulation of funerary practices in a given country: historically these have tended to evolve in accordance with the norms of historically dominant religions, and therefore may not be compatible with the religious practice of migrant minorities. In the following example from Denmark, the practice of grave re-use after a certain number of years provoked anxiety in an Iraqi respondent, for whom a burial plot in perpetuity is preferable:

> After 20 years [the Danish cemetery authorities] can take that grave … one time maybe, I will take my mother's grave [to Iraq] … take it to Iraq and bury her, because you have many rules over here which I wasn't satisfied with regarding that cemetery. (Iraqi, Female, 50s, first-generation, Denmark)

In contrast to this standpoint however a large proportion of Coptic Orthodox respondents expressed a high degree of indifference regarding burial location precisely from a desire to conform to the teachings of their church, notably as regards the nature of the soul and the body after death (summarised above):

> The body is ashes and dust so I don't mind where. [laughs] I look for, you know, for where my soul will be, that's the most important. (Coptic, Priest, 50s first-generation, UK)

> I live for God. Egypt is like Sweden when I die: the soul is going to God – my body, I don't think about it. (Coptic, Male, 40s, first-generation, Sweden)

Against Coptic indifference regarding the place of interment, a number of Iraqi respondents, exhibited a willingness to 'sacralise' space through their burial choices. The *sacralising* category expresses a preference based on the aspiration to symbolically and/or materially inscribe space with sacred meaning (see also Gardner 2002). This preference is usually expressed toward the country of residence, that is, the adopted homeland, although sacralisation through burial may also be directed at the country of origin, as the following quote shows:

> Actually for me, er, and even I have told all my family and friends, just in case, that I be buried in Iraq. That's very important. Because I love my monastery. And there is special places to monks and priests. They are still there with other - our previous priests - in a monastery near St Matthew's grave. And I don't know, I feel, I feel I belong to that area. The monastery's in the mountain, I belong to that area. (Iraqi, Priest, 30s, first-generation, UK).

The differences between Iraqi Christians and Copts concerning the sacredness of burial location are somewhat puzzling and not fully explicable from the interview data or from church eschatological doctrine, which as noted is rather similar across the different denominations. One may speculate however that the *sacralising* aspirations of Iraqi Christians are in some way connected to the greater existential threat which faces Iraqi Christianity at the present time. Leaving a material trace after death through burial in specific locations – either in Iraqi soil or in the lands of immigration – appears very intelligible when the legitimacy of lived religion is cast into doubt by destruction and dispersal.

Discussion: integration through burial?

As noted in the review of literature above, the decision to be buried in the new country of residence may be interpreted as a major reorientation of identity. It is a shift both in time and space, disaffiliating from the genealogical line of past generations in the place of origin in order to constitute the 'new first ancestors' buried in the adopted homeland, around whom future generations will congregate. As such, it has been argued that this shift in orientation constitutes the ultimate (in all senses of the word) marker of migrant integration. For Yassine Chaïb, discussing the final resting place of North African migrants and their descendants in France, 'the place of burial is the central geo-sociological element in the integration of immigrants to French society' (2000, p. 164; author's translation). Oliver (2004) also sees in the choice of burial a marker of integration and assimilation to the host society, in this case the integration of older British lifestyle migrants in southern Spain: 'Choosing burial in the cementerios demonstrate[s] commitment to Spain. It is a sign of assimilation or "going native", a clear marker that Spain is their home' (Oliver 2004, p. 249). In the remainder of this paper, I will critically evaluate this hypothesis of integration through burial. As will be shown, there are elements in our qualitative data which both support and question the idea that burial in the country of residence should always be considered as an indicator of integration. The nuance and ambivalence which emerges in interviewees' accounts shows the value of qualitative analysis in revealing the complexities of the relationship between place and identity over the lifecourse.

The term migrant integration likewise merits nuanced consideration. The term is commonly deployed to describe the process by which migrants' cultural and ethnic difference – what Modood calls 'post-migration difference' – ceases to be problematic in the context of the receiving society (Favell 1998, Modood 2012). A process of adaptation ensues: in more assimilatory contexts, the direction of adaptation is uni-directional, with migrants expected to adjust to the majority society with a minimum of disturbance to the latter. In more multicultural contexts accommodation is mutual and two-way, with majority institutions also recognising the social significance of migrants' group identities (Modood 2012). Given the broad scope of the concept, processes of integration take place in a very wide range of settings. Esser identifies four principal domains: cognitive (language, skills); structural (labour market participation, educational level, legal status); social (marriage, friendship, clubs, associations) and identificational (claims of belonging and identity) (Esser 1980, in Bommes 2012). It is in this latter identificational domain that the above-cited claims of integration through burial are best categorised.

Turning initially to evidence which questions the assumption of integration through burial, two points discussed above stand out. First, as was seen above, practical considerations such as the *security situation* in places of origin may inhibit the wish for a final resting place there, thus leaving the individual with little choice but to opt for burial in countries of immigration. It is therefore not a positively chosen active identification with an adopted homeland, but a decision which is constrained and provisional. *Religious conformity* is the second instance where burial in the country of residence does not necessarily equate with an integration perspective. Indeed, as was seen above in the case of the Iraqi woman who had deep misgivings about the practice of grave re-use in Denmark, and specifically insofar as it risked ontological security for her mother who is buried there, religious conformity can prompt individuals to exhume and repatriate their deceased loved ones. Similarly, for a large proportion of Coptic Orthodox respondents, burial in the country of residence does not constitute a deep sense of attachment or identity: rather, as was noted above, many Copts' testimony displayed a high degree of indifference as to where burial occurs.

Notwithstanding these narratives which denied integration through burial in countries of immigration, other interviews with Middle Eastern Christians gave strong backing to the contention. While sacred considerations, particularly conformity to church teachings, may impede a disposition towards integration through burial, they may also work in favour of interment in the adopted homeland. It was shown above that a 'sacralising' desire to make a particular place more holy was a strong rationale for an Iraqi priest to be buried near a monastery in Iraq. Another Iraqi priest, however, wished to sacralise a small corner of Britain which he had made his home:

> But I prefer to be buried in [the crypt of] my church, because I feel happy – why? Because I built it – it is my home. (Iraqi, Priest, first-generation, 50s, UK)

It was also noted above that the weight of the years spent in the country of residence, which I labelled *preponderant presence*, is also a factor in opting for burial there. 'For me personally to live two thirds of my life here and one third over there, this is where I belong' (Iraqi, Male, 60s, first-generation, UK). The size and concentration of a community of believers was also an indicator of 'preponderant presence'. The testimony of the young Assyrian/Syriac woman in Sweden mentioned above, to the effect that 'there are so many people here that [in death] they will feel home here too' shows how a sense of belonging is achieved over time and through the gathering together of the dead. A similar sense of belonging was aspired to by London's Assyrian community, who successfully lobbied the municipal authorities for a separate Assyrian section in a West London cemetery. In Denmark, however, where Middle Eastern Christian communities are less spatially concentrated, this effect did not materialise. These differences reveal the value in conducting a cross-country comparative analysis.

Conclusion

In a world where more and more people are physically mobile within and across national borders, the possibility to forge meaningful identifications with multiple places extends over the lifecourse, up to and even beyond the end of life. This opens up multiple options for where people envisage their final resting place to be. For migrant communities, especially those in which burial is the norm, the question of where to conduct funeral rituals may represent a stark existential choice: resolving this dilemma reveals much about how identities are negotiated in and through place.

This paper has explored this relationship between identity and place through the prism of preferred burial location. The contributions of the paper have been threefold, respectively, in terms of methodology, theory-testing and analysis. The paper's first contribution has been to synthesise the few existing sources of literature on this topic, primarily qualitative but also quantitative: sources which, by and large, have not been in dialogue with each other thus far. In addition to establishing the current state of the art, I have developed a new empirical strand in this literature beyond the predominant focus on Muslim communities in Europe, by studying the question of preferred burial location in Middle Eastern Christian migrant communities. Second, I have sought theoretical clarity by examining the hypothesis that burial in the country of residence constitutes a straightforward indicator of migrant integration. Last but not least, my third aspiration with the paper was to aid future reflection and analytical clarity by developing a typology of motivations for different burial location preferences.

In this latter regard, various sets of rationales for preferred burial location were elaborated. The analysis of the data first of all puts centre stage the practical considerations which influence burial location, namely financial and organisational costs, as well as security concerns. These practical considerations have been underplayed in the existing literature to date. I then moved to elaborate further categories of importance which emerged in the interview data, drawing on existing studies which have given priority to family, territorial and religious considerations. When discussing family factors, it was shown that temporal orientations are paramount: respondents were variously oriented to a past perspective, via forebears in the place of origin and deceased parents; others gave priority to 'significant others' in the present, while a third position was to construct an identity of 'new first ancestor' in the country of immigration, around whom future generations would gather. Territorial considerations stressed the weight of the presence of a migrant community in a given location, in quantitative and qualitative terms. Others evoked allegiances to the nation, or emotional connections to a particular place or landscape, as rationales for preferring one burial location over another. Sacred considerations were also a strong feature of the qualitative data, and a number of respondents expressed a desire to 'sacralise' space through their burial choices, that is, to symbolically and/or materially inscribe space with sacred meaning. The concern to conform to religious orthopraxy was a key finding, particularly for Copts, for whom there was a high degree of religiously justified indifference to the question of burial location.

This indifference speaks to a broader question in the literature, namely whether we can consider burial in the country of residence as an ultimate indicator of integration (Chaïb 2000, Oliver 2004). There were elements in the data which both supported and challenged this hypothesis. Given the dire security situation for religious minorities, including Christians, in several parts of the Middle East, burial in Europe is not necessarily always a positively chosen active identification, but rather a decision which is constrained and provisional. Interestingly, this ambivalent finding on integration finds an echo in another contested literature, namely the question of the sustainability of diaspora communities. When the full range of lifecycle-based rites of passage, up to and including death rituals, can be enacted in countries of immigration, then the rationale for return to the 'homeland' becomes less self-evident (Hunter 2015). If it is accepted that diasporas are predicated in part on this desire for eventual return (Safran 1991), then it follows that the enactment of these lifecycle rituals in countries of immigration may lead to the dissolution of the very ties of diaspora which they purport to uphold.

Notes

1. One exception is Gunaratnam (2013), who gives the example of a Kenyan man diagnosed as HIV positive whose wish to be repatriated from Britain to Kenya for burial could not be fulfilled due to the administrative difficulties of repatriating bodies with HIV infection.
2. The vast majority of our interviews took place between February 2014 and July 2014, that is, just before the advance of the Islamic State Organisation (also known as ISIS) through Northern Iraq. It may be speculated that security would have been mentioned by our respondents even more had we conducted interviews after July 2014. Numerous acts of desecration of Christian graves have been reported since the organisation's occupation of large swathes of Northern Iraq. http://www.dailymail.co.uk/news/article-3043255/ISIS-destroy-Christian-graves-headstones-sledgehammers-Islamist-terror-group-continues-purge-against-religions.html.

Acknowledgements

I am particularly grateful for the support of my colleagues Fiona McCallum, Lise Paulsen Galal, Marta Wozniak-Bobinska and Sara Lei Sparre in carrying out interviews for this research. I am equally grateful to Anne Rosenlund Jørgensen for translating interview material from Danish into English. This article has also benefited greatly from the insights of Eva Soom Ammann and two anonymous reviewers.

Disclosure statement

No potential conflict of interest was reported by the author.

Funding

The DIMECCE project is funded by the European Union's Seventh Framework Programme for research, technological development and demonstration under grant agreement no 291827. The project Defining and Identifying Middle Eastern Christian Communities in Europe is financially supported by the HERA Joint Research Programme (www.heranet.info) which is co-funded by AHRC, AKA, BMBF via PT-DLR, DASTI, ETAG, FCT, FNR, FNRS, FWF, FWO, HAZU, IRC, LMT, MHEST, NWO, NCN, RANNÍS, RCN, VR and The European Community FP7 2007-2013, under the Socio-economic Sciences and Humanities programme. This article was written during a visiting fellowship held at the Institut national d'études démographiques and generously funded by the Ville de Paris.

References

Ansari, H., 2007. 'Burying the dead': making Muslim space in Britain. *Historical research*, 80 (210), 545–566.

Attias-Donfut, C. and Wolff, F.C., 2005. The preferred burial location of persons born outside France. *Population (English edition)*, 605 (6), 699–720.

Bommes, M., 2012. Transnationalism or assimilation? *In*: C. Boswell and G. D'Amato, eds. *Immigration and social systems: collected essays of Michael Bommes*. Amsterdam: University of Amsterdam Press, 107–124.

Casal, A., Aragonés, J.I., and Moser, G., 2010. Attachment forever: environmental and social dimensions, temporal perspective, and choice of one's last resting place. *Environment and behavior*, 42 (6), 765–778.

Chaïb, Y., 2000. *L'émigré et la Mort: la mort musulmane en France*. Aix-en-Provence: Edisud.

Clayton, J., 2009. Thinking spatially: towards an everyday understanding of inter-ethnic relations. *Social and cultural geography*, 10 (4), 481–498.

Cody, A., 1991. Eschatology. *In*: A. Atiya, ed. *Coptic encyclopedia* Vol. 2. New York: Macmillan, 973a–974b.

Cresswell, T., 2004. *Place: a short introduction*. Malden, MA: Blackwell.

Esser, H., 1980. *Aspekte der Wanderungssoziologie*. Darmstad: Luchterhand.

Favell, A., 1998. *Philosophies of integration : immigration and the idea of citizenship in France and Britain*. Basingstoke: Palgrave.

Galal, L., *et al.*, 2016. Middle Eastern Christian spaces in Europe: multi-sited and super-diverse. *Journal of religion in Europe*. doi:10.1163/18748929-00901002.

Gardner, K., 2002. *Age, narrative and migration: the life course and life histories of Bengali elders in London*. Oxford: Berg.

Gunaratnam, Y., 2013. *Death and the migrant: bodies, borders and care*. London: Bloomsbury.

Hallam, E. and Hockey, J., 2001. *Death, memory, and material culture*. Oxford: Berg.

Hunter, A., 2015. Deathscapes in diaspora: contesting space and negotiating home in contexts of post-migration diversity. *Social and cultural geography*. doi:10.1080/14649365.2015.1059472

Jassal, L., 2015. Necromobilities: the multi-sited geographies of death and disposal in a mobile world. *Mobilities*, 10 (3), 486–509.

Jonker, G., 1996. The knife's edge: Muslim burial in the diaspora. *Mortality*, 1 (1), 27–43.

Kaplan, D.H. and Chacko, E., 2015. Placing immigrant identities. *Journal of cultural geography*, 32 (1), 129–138.

Maddrell, A. and Sidaway, J.D., 2010. Introduction: bringing a spatial lens to death, dying, mourning and remembrance. *In*: A. Maddrell and J.D. Sidaway, eds. *Deathscapes: spaces for death, dying, mourning and remembrance*. Farnham: Ashgate, 1–16.
Marjavaara, R., 2012. The final trip: post-mortal mobility in Sweden. *Mortality*, 17 (3), 256–275.
Modood, T., 2012. *Post-immigration 'difference' and integration: the case of Muslims in Western Europe*. London: British Academy.
Oliver, C., 2004. Cultural influence in migrants' negotiation of death. the case of retired migrants in Spain. *Mortality*, 9 (3), 235–254.
Proshansky, H.M. Fabian, A.K., and Kaminoff, R., 1983. Place-identity: physical world socialization of the self. *Journal of environmental psychology*, 3 (1), 57–83.
Rallu, J.-L., 2016. Projections of older immigrants in France, 2008–2028. *Population, space and place*. doi:10.1002/psp.2012
Reimers, E., 1999. Death and identity: graves and funerals as cultural communication. *Mortality*, 4 (2), 147–166.
Rowles, G. and Comeaux, M., 1987. A final journey: post-death removal of human remains. *Tijdschrift Voor Economische En Sociale Geografie*, 78 (2), 114–124.
Safran, W., 1991. Diasporas in modern societies: myths of homeland and return. *Diaspora: a journal of transnational studies*, 11, 83–99.
Tan, D., 1998. *Das fremde Sterben*. Frankfurt am Main: IKO-Verlag.
Teather, E.K., 2001. The case of the disorderly graves: contemporary deathscapes in Guangzhou. *Social and cultural geography*, 2 (2), 185–202.
Tonkinson, M., 2008. Solidarity in shared loss: death-related observances among the Martu of the western desert. *In*: K. Glaskin, M. Tonkinson, Y. Musharbash and V. Burbank, eds. *Mortality, mourning and mortuary practices in indigenous Australia*. Farnham: Ashgate, 37–53.
Venhorst, C., 2013. *Muslims ritualising death in the Netherlands: death rites in a small town context*. Münster: LIT Verlag.
Wissa-Wassef, C., 1991. Funerary customs. *In*: A. Atiya, ed. *Coptic encyclopedia* Vol. 4. New York: Macmillan, 1124a–1125b.

Uncertain Belongings: Absent Mourning, Burial, and Post-mortem Repatriations at the External Border of the EU in Spain

Gerhild Perl

Institute of Social Anthropology, University of Bern, Bern, Switzerland

ABSTRACT
Since the mid-1980s, migrants from North African and sub-Saharan countries have irregularly crossed the Strait of Gibraltar in the hope of a better future for themselves and their families. Travelling in small, poorly equipped boats without experienced captains has cost the lives of myriad border-crossers. Exploring the junction of death and belonging, I first open a discussion about the enigmatic relation between a dead body and a dead person and argue for the importance of the physical presence of the body for mourning. Second, I show how the anonymity of dead border-crossers and their uncertain belongings are generated, concealed, or rewritten. Following the story of an undertaker, I third examine post-mortem border crossings. Depicting the power relations within identification processes, I outline the ambiguity of the term belonging by emphasising the constitutive significance of personal belongings such as clothes to restore a person's identity. Reflecting on the ethical relationships which different actors (including the researcher) undertake with the deceased, I aim at gaining a better understanding of the multiple belongings of dead border-crossers found on Spanish shores.

A white wall in-between recesses, Spanish graves situated side-by-side and one above the other. They are called *nichos*: above-ground alcoves decorated with artificial flowers and covered by headstones. Most of them say a name, a date, and an epitaph. I walk up and down the paths between the walls where the coffins are inserted, searching for hints of the presumed final resting places of deceased border-crossers at cemeteries on the Spanish Side of the Strait of Gibraltar. So far, I have found *nichos* with different inscriptions: *Inmigrante de Marruecos* (immigrant from Morocco), *desconocido* (unknown), and *restos* (remains). Some *nichos* simply have a cross, accompanied by a date and a number, scratched into toxic fibre cement; some have scarcely decipherable judicial abbreviations written on bare surfaces with pale green chalk, which easily fades. And other burial sites are not even identifiable as graves. I stand in front of a white, neutral, blank wall. If a cemetery worker had not told me, I would never have imagined that 10 embalmed bodies are buried behind it. They are victims from a shipwreck in 2003 at the Western coast of

Andalusia. They should have been repatriated but due to difficulties in the identification process the mortal remains are still in Spain.

The cemeteries do not have a standardised way of marking the graves of the unknown, and the inscriptions differ widely – as do estimates of the numbers of the dead. It is almost impossible to evaluate how many people have died since irregular migration to Europe increased in the late 1980s. Available tallies vary greatly and should be approached with caution. Last and Spijkerboer (2014, p. 85) argue that the variation in the number of border-related deaths results from the different interests pursued by different actors and on the paucity of available data from the maritime border zones of the EU. The number of unauthorised crossings – and by extension, deaths – at the *Estrecho* (Strait of Gibraltar) increased after Spain entered the Schengen Area in 1991 and introduced visa requirements for Moroccan citizens (Alscher 2005, Ferrer-Gallardo 2008, Last and Spijkerboer 2014). Since the installation of the national surveillance system SIVE (*SistemaIntegrado de Vigilancia Exterior*) between 2003 and 2008, and bilateral agreements between Spain and Morocco as well as West African countries, crossings became more difficult (Ferrer-Gallardo 2008, Andersson 2014). Yet, tightened border control does not lead to a decrease in illegalised border crossings or the prevention of death but instead results in the displacement of migratory routes and in an increasing number of people who die[1] (Grant 2011, p. 140).

In this article, I attempt to disentangle the multifaceted and ambiguous be/longing, actions, and politics that emerge in the treatment of dead bodies. To understand belonging after death, I first question how we can think about the concrete presence of dead bodies and their rather nebulous absence. I argue that the absence of bodies evokes uncertainties about the truth, circumstances, and whereabouts of the death of a person and forecloses grieving, whereas their physical presence provides answers and thus has the potential to relieve and pacify the bereaved. Further, I reflect on the ethics of research and writing in this area. Second, I emphasise the actual whereabouts of the bodies by exploring how their national and religious belongings are displayed and visualised at Spanish cemeteries. Third, I discuss post-mortem repatriations as a complex, contested, and contradictory social field. Following the story of an undertaker, I examine the extent to which border-related deaths influence and shape feelings and actions of local people living in the border area and I draw a comparison between the personal engagement of the undertaker and the ethnographer. Finally, I conclude that the analysis of multiple and uncertain belongings of the dead, which I discuss throughout the article, needs to form part of the ethical demand for identification.

For my research[2] I have conducted fieldwork in Southern Spain and I have undertaken extended field trips to Morocco in 2012, 2014, and 2015. I have traced and followed the paths of the deceased and the stories about them through multi-sited ethnography (Marcus 1995). The core of my methodological approach has encompassed participant observation, informal talks, and semi-structured interviews with border-crossers, cemetery workers, undertakers, police officers, NGO members, judges, lawyers, local politicians, human rights and Christian activists, representatives of Islamic communities, people who live in the border area, and families of deceased border-crossers. It is important to note that most of the interviewees preferred to be identified by their real names (sometimes with, sometimes without surname) and thus become visible in their professions, evaluations, feelings, and thoughts. If not specified otherwise, their real names are used in the article.

Present bodies and the *absence of the absence*

Living in transnational settings, migrants have relations beyond national, geographical, and cultural borders and boundaries. Mobility, interconnectedness, and multiple belongings in various social fields shape their everyday practices (Basch *et al.* 1994, p. 7, Strasser 2009a, 2009b, p. 74). To understand the complexity of multiple belongings and the specific forms of networking and intervention, the social anthropologist Sabine Strasser suggests 'using the notion of belonging as a mode of thinking about how people *get al*ong and how various forms of being and longing are articulated' (2009c, p. 187; see also Probyn 1996, p. 5). Further, Strasser (2009a: 31–32) points out that *belonging* signifies both, *being* and *longing*. While the former refers to the 'concrete presence and experience' of a person, the latter expresses 'desire and imagination'.

The term *being* is a complex and challenging word to think within the context of a person's death. The dead are present as corpses and yet they *are* not anymore. Corpses have eyes, but they don't look at us, they have mouths, but remain voiceless, they have hands, but they don't touch – in other words, corpses seem to neglect any relation with the living, as the philosopher Macho (1987) suggests when he defines death as the irrevocable rupture of communication. Macho (1987, p. 409) speaks about a 'conundrum of the corpses' to comprehend their intransigent resistance towards any kind of social obligation and to sketch the 'inexplicable duplication', which consists in the fact that a dead body *is* and *is not* identical with the dead person. Corpses and their very concrete presence represent the former person as 'the presence of an absence' (Macho 2000, p. 99). The dead *is* and *is not* at the same time.

Macho's phenomenological approach helps to understand the unsettling, inscrutable, and ambiguous character of corpses and sheds light on the puzzling junction between the dead body and the dead person. But exploring the quandaries between belonging and death, I prefer to think of the dead body as constitutive of social relations, rather than indicative of their absence. Inspired by the social anthropologist Magaña (2011, pp. 159–160) and her contention that a dead body and its display contain a 'performative quality' and her idea 'that in death, the […] body bears the possibility to resignify social, political, and spatial relations', I am interested in how different actors relate ethically to the presence of a dead, unknown body. Therefore, the analysis of people's speech becomes crucial. The anthropologist Jackson (2013, p. 4) uses the notion 'arresting images' to point out an *ontologising* tendency in migration studies when addressing (living) migrants as a collective 'them'. This form of 'social and discursive violence […] reduces the other to a mere object – a drudge, a victim, a number' (Jackson 2013, p. 5). In the context of my fieldwork, public discourses, academic research and writing as well as day-to-day speech often refer to unidentified dead migrants as anonymous corpses or cadavers – a discursive practice that disguises the dead person as well as his or her belonging and reduces the dead to a mere body – a causality, a fatality, a statistic. In my view, exploring the belonging of a dead body forces us to think of the body as the representation of the former person rather than as a mere object. The phenomenologist Hasse (2016)[3] argues that the objectifying significance of the terms 'corpse' *(Leiche)* and 'cadaver' *(Kadaver)* creates not only an 'emotional distance' but also denies the personal history and dignity of the dead.

A striking example that follows this objectifying practice is an interdisciplinary study with a strong forensic emphasis carried out in the US–Mexican borderlands. The study aimed at documenting the impact of the local ecology for 'the process of corpse decomposition and taphonomy' (Beck *et al.* 2015, p. 19). To do so, three pigs 'dressed in clothes similar to those worn by migrants' (Beck *et al.* 2015, p. 12) are killed and left in the Sonoran desert to study the changing state of the cadavers. In this way the researchers draw a comparison with the physical fate of dead border-crossers: such an investigation is characterised by great emotional distance and driven by a scientific urge to rationalise death. Not only is the killing of animals ethically questionable, so too is the researchers' disregard for personhood and belonging in death: the death of a person becomes the same as the death of a (killed) animal. In contrast, other studies dealing with the fate of dead bodies in the Sonoran desert emphasise migrants' personhood. To cite just a few, Regan (2010) focuses on the lived experiences of the bereaved, border-crossers, activists, and state actors; Reinecke (2013) draws attention to the impact of complex and excluding identification procedures for families; and Magaña (2008, 2011) examines the 'political afterlife of dead bodies'.

Being confronted with border-related death, it is important to remember those who have disappeared: they have friends and families who miss them, who wonder about their whereabouts, who hope that they are still alive, and who long to know what has happened to them. In Morocco an *Association of Friends and Families of the Victims of Clandestine Immigration* (AFVIC) has been founded and the monthly television programme *Moukhtafoun* (The Disappeared) announces details of missing persons in order to find them. Tunisian families whose relatives went missing in the attempt to cross the border to Italy have also formed an association demanding to know what exactly happened to their loved ones (Moorehead 2014). These initiatives show how people try to *get along* with the uncertainty of death and how the *longing* for knowledge is articulated (Strasser 2009c, p. 187). But most families remain in a state of not knowing what happened to their relatives and not knowing how to demand the right to know.

Grant (2011, p. 142) points out that basic human rights are clearly violated at the EU's external borders. These rights – established in international humanitarian law as formulated by the UN Human Rights Council – include: restoring the identity of the deceased, the 'clarification of the fate of a missing person', the duty to inform families, the 'restoration of family links', and the repatriation of mortal remains. These rights however are not applied to illegalised migrants (2011, p. 143).

In the light of the uncertainty of a person's fate and belonging, I reconfigure Macho's notion of the 'presence of the absence' as the *absence of the absence*. Many border-crossers are lost at sea and their mortal remains will never be found, others are washed up at the shores or discovered in the water. As I have noticed during my fieldwork, although DNA-samples are taken from these bodies, the families of the deceased are rarely informed of the death. The dead remain missing; uncertainties and fears survive and may agonise the living. A dead body materialises not only the absence of the former person, but proves the certainty of death. Being deprived of the body and/or the knowledge of its demise keeps the uncertainty of death alive, and therefore the *absence of the absence* means the *absence of the certainty of death.*

If families are given notice but cannot see, touch, wash, or feel the body, they experience the concrete absence of the body and this influences grieving and mourning processes tremendously: they do not only grieve the demise, but also the loss of the mortal remains and the non-realisation of appropriate practices (Weiss-Krejci 2013, p. 290). The absent body becomes the subject of longing and desire. The philosopher Butler (2004, p. xiv) identifies a 'differential allocation of grievability' that make some lives worth protecting, saving, and mourning, while other lives remain unacknowledged, unprotected, unremembered, and ungrieved and thus, they are not apprehended as living in the first place. Butler's observation offers the possibility to consider grief not only as a personal matter and private feeling but also as a phenomenon that helps to understand the making and governing of precarious lives (and deaths). The political dimension of public mourning could give back personhood to the deceased in public space and discourses.

Recovering, identifying, informing, confirming, and repatriating are fundamental procedures to bring closure to families and friends. Knowing about the death of a person is not only necessary for mourning and pacifying death but is also important for going on with life. Additionally, the issuing of a death certificate may be a necessary legal ground for inheritance and re-marriage (Grant 2011, p. 146). Yet, most of these bodies remain unidentified and thus their families are often condemned to a state of not knowing. Emphasising the whereabouts of these bodies at local cemeteries in Spain, I will now examine the intersection of burial practices with national, genealogical as well as religious belongings.

Buried the Spanish way

The research for this article was conducted in the Spanish border town of Algeciras – a name derived from Arabic Al-Jeezira al-Khadra (the green island) – and the Campo de Gibraltar, the town's surrounding area. People who live in the area remember with sorrow the many border-crossers who drowned in their waters. However, this was in the past: 'Today it does not happen anymore, not so much, this was at the end of the 1990s until 2005 or 2006, maybe, today it happens in Italy', says a cemetery gardener who I met at the old graveyard in Algeciras. Most of the people I talk to are well aware of changing migratory routes. Cemetery workers, especially, notice a shift: 'It is not like before, before every week, every day we buried an immigrant who drowned' (Diego[4]). Several of these dead from 'before' are buried at local municipal or Catholic cemeteries close to the place of recovery, where inscriptions, filing, book-keeping, and ways of commemorating differ greatly.

Kisko, a gravedigger, takes me around the Catholic cemetery of Tarifa. We stop in front of two *nichos*. Both of them say in black letters on white stone: *Inmigrantede Marruecos 2009*. Kisko looks at me and says that he buried these two men. He assures me: 'They were not from Morocco'. 'But why does it say immigrant from Morocco?', I ask. He shrugs his shoulders: 'The headstones are standardised'. Both bodies are unidentified: neither their nationality nor their creed is known. Kisko's assurance that they are not Moroccans is based on his observations prior to burial of their physical appearance 'which was not Arabic'. Furthermore, he explains that in 2009 Moroccans no longer crossed the *Estrecho* in boats. Indeed, nowadays most border-crossers originate from sub-Saharan countries and therefore it is easily possible that the two dead men were not Moroccan.

Leaving the cemetery I began to wonder if the inscription *Inmigrante de Marruecos* solely indicates a Moroccan nationality or if it could perhaps refer to sub-Saharan migrants who took the boat in Morocco to cross to Spain. When I showed the photo of the headstones to Spanish native speakers, they consistently agreed on Morocco as the national identity of the dead. Thus the headstones inscribe a misleading national belonging of the dead. And since the great majority of the Moroccan population is Muslim, the inscriptions also imply an association with the Islamic faith, which may or may not be true. According to Kisko, a small Christian organisation organised and paid for the 'standardised headstones', yet, he could not say why they did not organise new headstones. The anthropologist Verdery (1999, p. 29) stresses the speechlessness of dead bodies, which allows the living to rewrite history and to use dead bodies as (political) symbols by emphasising selective parts of their stories and by putting words into their mouths. At the cemetery in Tarifa, the story of those two men has been rewritten by assigning a national identity which is highly doubtful and by insinuating a religious belonging which is in fact unknown.

When I phoned the person who initiated the headstones and asked him for a meeting, he refused to talk to me: 'It is a private matter', he said and hung up. Yet, death at the border and the ways of commemorating it are not private but rather political matters. As Magaña (2008, p. 108) states, border-related deaths are not merely private losses because their recovery, identification, and the ways they are made visible or not in public space enable or foreclose not just mourning, but also state recognition and political claims. Although I do not know why the Christian organisation decided on these two headstones, I assume that they wanted the death of those two men to be recognised and acknowledged. But even if their intention was to make them visible in public space and to commemorate them, the way they did so is problematic. To affix headstones suggesting a national identity which in fact is unknown underlines the precarious and extremely disenfranchised status of (dead) unidentified migrants. Yet, one should keep in mind that people who live in the area and who do not want to ignore the migratory deaths in their midst are overburdened with the question of how to proceed properly with the dead. In talks with cemetery workers, a strong emphasis on the equal treatment of dead migrants and anybody else is often expressed. According to Diego, who has been working at the cemetery in Algeciras for 15 years:

> How they die, it is sad, very sad, but there is also something that draws my attention ... the good in all this bad ... you treat everybody equally ... everybody receives the same examinations, the same treatment, the same autopsy, the same professionalism, the same effort. An immigrant who dies drowning in the *Estrecho* receives the same treatment like a multimillionaire who drowns during surfing in Tarifa. And both die in the *Estrecho*, that's curious ... The circumstances of death are different but the professionalism is the same, the expenses are the same.

Diego stresses the consistent equality as politically correct and morally permissible, and indeed, at first it was comforting to hear him say that. But a second look reveals that this equality evokes conflicted meanings, since it is determined by national, cultural, and religious norms and regimes in the receiving country: it seems to me that to 'treat everybody equally' means that in most of the cases everybody is treated either as an atheist or a Catholic. People who never lived but died in Spain usually do not have anybody who is morally qualified to decide on the how, when, and where of the burial and thus they are subjected to dominant local burial norms. Restrictive visa policies prevent families to come to identify the body, to perform the burial, or to repatriate the dead. Furthermore, survivors who travelled with the deceased often neither report a missing person nor identify a body because they fear deportation or other consequences of their 'immigration position' (Grant 2015, p. 12).

Both the rewriting of a dead person's identity as well as the claim for equal treatment obscure not only the uncertain and diverse belongings of the dead but also social inequalities produced by the EU's immigration and border policies. The continued policy of closure prevents people from entering the Schengen territory legally and forces them to take life-threatening routes to bypass controls. Drowning, hyperthermia, and suffocation are generally considered to be the main causes of death at sea. Yet, following Magaña (2008, p. 110), the pronouncement of natural causes of death forms part of the 'state's attempt to delimit understandings of violence around these deaths [...] displac[ing] the possibility of culpability outside of the state itself'. The state rejects not only culpability and accountability for the circumstances of death, but also refuses to take responsibility for the fate of the deceased.

In conversations with people who had crossed the border many reported that faith forms a profound part of their life. Faith is not only a source of strength, consolation, and hope but also indicates a person's belonging. Since religious denomination is not determined and not always possible to determine, the question of how to bury a person properly and in a dignified way is overwhelming for local actors. Not just undertakers and cemetery workers but also local politicians complain that the Spanish Government leaves them alone with the treatment of the dead, especially those who are Muslims. Even if religious belonging of the deceased were to be known and if somebody were to demand a respectful burial, the question of how and where to bury deceased Muslims properly is a contested issue in the Campo de Gibraltar area. In 1992, Spain approved a law that imposes the concession of separate sections for Muslims in municipal cemeteries and in Andalusia inhumation without a coffin has been possible since 2001 (Tarrés *et al.* 2012, pp. 7–9). Yet, the Campo de Gibraltar area is not only devoid of an Islamic cemetery but also of separate sections for Muslims in municipal cemeteries. Visiting the new and non-confessional municipal cemetery in Algeciras, I noticed a strong dominance of Catholic burial traditions. Furthermore, the cemetery does not offer specific Islamic burial services: inhumation without a coffin is impossible, graves are not facing towards Mecca, and they are not perpetual so may be re-used. Representatives of Islamic Communities in Algeciras state that Muslims with Moroccan background either bury their dead at the private Islamic cemetery, 70 miles East of Algeciras in Fuengirola, where Islamic burial practices are accomplished, or they transfer them to Morocco.

There exists a significant diversity regarding proper Muslim burial practices in Europe among scholars and also among members of Islamic communities. According to the Islamic scholar Habib Rauf (2014: 34–37), a burial in a Muslim country is not required by Islamic teaching but the fulfilment of Islamic burial practices is; furthermore, post-mortem repatriations contradict the Qur'an. Yet, Omar Samaoli, a gerontologist who has paid significant attention to the end-of-life practices of North African migrants in France (2007, pp. 129–130), notes the permissibility of post-mortem repatriations in the jurisprudence of the Maliki rite – one of the four major legal traditions in Sunni

Islam – which is dominant in Morocco. In conversations I have had with Moroccan families from the Middle Atlas whose deceased relatives are buried in Spain, the desire to 'bring them home' was striking. It seems that the longing for repatriation is motivated more by emotional and cultural be/longings than by religious ones, and can be understood as a form of 'genealogical reconstruction' (Tarrés *et al.* 2015: 9–12) that pacifies the bereaved and reinstalls belonging after death.

But families do not always desire to repatriate the dead. Studies concerned with the fate of deceased border-crossers in the Sonoran desert show that families who had immigrated to the USA prefer to bury the dead in US-American soil (Regan 2010, p. xviii, Reinecke 2013). In transnational settings the question of to whom a dead body belongs is crucial. Families often articulate the desire to bury the dead close to them; they want to be able to visit the grave and to mourn the deceased. But in situations where post-mortem repatriations or other transnational post-mortem transfers are not accomplished by the governments concerned – be that in the countries of origin, transit or destination – deceased border-crossers are buried close to the place of recovery. In Spain this generally means they are subjected to hegemonic local burial norms and are literally incorporated in the (undesired) receiving country. Yet, for a minority of Moroccan border-crossers, repatriation has been effected thanks to the initiative of a local undertaker, as I will discuss in the following section.

Between business and engagement

Against the narrative presented above, one person who does take religious belonging into account is the Spanish undertaker Martín Zamora. Focusing on his story, I will discuss the power relations inherent in identification processes. In so doing, I attempt to elucidate the ambiguity of the term 'uncertain belongings' and show how border-related deaths have influenced and shaped the work, feelings, and thoughts of the undertaker.

I met Martín Zamora in Ceuta, the autonomous Spanish enclave on the African continent, where he is building a new funeral parlour with the intention of providing undertaker services to the Muslim population. Even though Ceuta has an Islamic cemetery and a large Muslim population, no funeral parlour offers specifically Islamic services. Martín Zamora has known the funeral business since he was 14 years old. Coming from Murcia, an autonomous region in the Southeast of Spain, he settled down in Los Barrios, a small town next to Algeciras and opened a funeral parlour, a so-called *tanatorio* with its own morgue. This was at the end of 1998 when the *Estrecho* was still one of the most important migration routes towards Europe.

Until he picked up the first bodies in 1999, he was not aware of border-related deaths in Spain, as he recalls:

> I had a lot of questions … these were persons without identification and they (*the authorities*) told me that they should be buried in the area where they were recovered. I found it very sad, persons who must have a name … family who certainly wants to know something about them. I wanted to do something … locate the family and give a name to the deceased … I wanted to identify them.

Martín Zamora had not expected the sudden confrontation with the deaths of dozens of unknown border-crossers. But his empathy with the survivors was not the only motivation for identifying and repatriating them. At first, his motivation was both compassionate and calculating:

> As a businessman, let's be realistic, this is the first thought that comes into mind. Later it became another thing, but in the beginning, it was this ... if I have twenty bodies here and I succeed in identifying them, somebody will pay for the transfer of the dead ... of course, later you notice that these transfers you have to pay, all of it. Because the families did not have the money and nobody wanted to know anything about it, even if you have located a father and you tell him, look I have your son, and you tell him that this costs 3000 or 3500 Euro and he doesn't have the money ... you can't say, I'm sorry, but I won't bring your son, he will be buried in Algeciras. So, what we did a lot of times, we put in the money, we paid all the expenses, and we repatriated the dead.

Martín Zamora and his co-workers not only repatriated people whose families could not afford the expenses. Over time, many Moroccans got to know and to trust him, which in turn was good for his business. Furthermore, he hired Arabic-speaking staff, offered Islamic burial services, and built a mosque in the *tanatorio*. Eighty to ninety per cent of his everyday business consisted in the treatment and repatriation of deceased migrants of Moroccan origin. But when the economic crisis hit Spain many migrants and possible clients left and thus his business suffered and the money for the free repatriations dried up. In 2013, he finally sold the *tanatorio*.

Martín Zamora made an exceptional effort to identify deceased border-crossers. The first pre-condition for repatriating a dead body and for clarifying a dead person's belonging is to know the name and the place of residency of the deceased and the family. Yet, as Grant (2011: 147–148) stipulates, one consequence of illegalised border crossing consists in the fact that many migrants travel without identity documents or use forged papers and thus their deaths often remain anonymous. Furthermore, police do not make enough efforts to identify an unknown border-crosser. As Martín Zamora puts it:

> We were confronted with the handicap that nobody dealt with the problem ... the courts, the Spanish police – well, those who actually should worry about finding the families, they didn't do it. So what I did was, I tried to compile all the information I could get and I searched the personal belongings of the dead ... I tell you, as we became more experienced we figured out that the dead often had photocopies sewn in their clothes, but above all they had some phone number of a friend or relative in Europe ... so we began to examine the dead more carefully, let's say, we made a more exhaustive search of the clothes and we collected all the information we could get ... well, we always tried to identify the dead, yes.

In Spain, the forensic departments of the police and the Guardia Civil are in charge of identification and Interpol is responsible for communicating with the country of origin in order to exchange data. Recently, Spanish police seem to be making more thorough efforts to identify a dead person. A policeman from the forensic department in Madrid stated that his colleagues needed to 'try harder' to identify deceased border-crossers, on account of the war in Syria since 2011. This was because, in his view, many 'educated refugees who have money' try to cross the border to Spain and he is convinced that the families of those who have died will make a fuss in the future: 'They will ask for answers'. For him, the duty to identify derives from the possible problems, which 'educated and wealthy refugees' might cause in the future. This statement contrasts with most of the declarations of other police officers I spoke to. Most of them stress that markers of differentiation such as the skin colour of a person, class, the country of origin, and legal status do not influence their work in any way. Since 1999, the extraction of DNA-samples forms a standardised procedure and is considered the most reliable source for identifying unknown bodies. But solely extracting DNA from the corpses without comparing it to relatives' DNA does not lead to successful identification. Although I am aware of one case in which Spanish and Moroccan authorities collaborated to extract and compare DNA-samples, cross border agreements and procedures to identify and return the dead are not developed. This contrasts with the well-established cooperation between Spain and Morocco as well as Spain and West African countries to tighten border control and to prevent people from crossing (Andersson 2014, Last and Spijkerboer 2014). While a lot of effort is expended to identify and potentially repatriate undocumented, living migrants, dead border-crossers remain largely unidentified and thus not repatriated (Zagaria 2011).

Martín Zamora points to emotional and moral difficulties that emerge when authorities mainly rely on DNA-samples. Following his experiences, some judges only issue approvals for repatriation if the DNA of the body matches with a relative, and they ignore identifications based on the recognition of the body or the clothes. He remembers a drowned border-crosser from Morocco who was 'completely identified' by his brothers, who were Moroccan with an Italian nationality.

> The brothers identified the body, but the judge did not approve ... but the corpse was so characteristic because an arm was missing ... but the DNA was negative ... the judge never authorised the repatriation, we had to put up with it and bury him in Los Barrios. In other words, there were cases with failures regarding the DNA, but we do not know if this was an administrative failure or a mistake, or maybe he was not the biological brother. But for them, he was their brother.

The question where and to whom this man belongs is not uncertain at all. Not trusting the word of the brothers, but intransigently relying on genetics – even if the possibility of a mix-up is known – demonstrates how biopolitical power regulates, manages, and organises the fate of a body. Biomedicine separates the self from the body (Sharp 2009, p. 290) and thus the body is reduced to a biological object, which becomes completely detached from the person. Additionally, the arbitrary decision of the judge and his powerful position underlines the disenfranchised status of the dead migrant and his relatives. The one-armed man and nine others are victims from a shipwreck in 2003 and they are buried behind the white, neutral, empty wall, which I mentioned in the introduction. Their graves are not recognisable as burial sites: nothing indicates that somebody is resting there. Albeit in vain, the brothers were able to come to Spain to try to identify the deceased, thanks to their Italian nationality.

Dependent on visa policies and the benevolence of the judges, a body may or may not be repatriated. Fortunately, other judges have supported Martín Zamora's endeavour and allowed him to take clothes and personal belongings of deceased border-crossers to Morocco, in order to find their families. He, his brother, and co-workers went to the market places of the villages from which they thought the deceased had come from. They exhibited the clothes and other belongings the dead carried with them and waited for somebody to recognise the things. In this way they could identify most of the deceased.

Spanish authorities often do not acknowledge the value of the personal belongings for identification processes and their main approach to identifying a person is based on DNA. Yet, Martín Zamora's examination of the deceased's clothes, his struggle to bring their belongings back to Morocco in order to trace their origins and identities exemplifies the contested identity of the dead body. Authorities seek to reveal the identity and the national belonging due to genetic procedures performed on corpses. However, the belongings of the dead – their personal effects, clothes, assets, and possessions they travel with – reveal the dead person. In other words, a corpse becomes a dead person who belongs to somebody and thus the personal belongings of the dead reconstitute the national, religious, and genealogical belonging of the dead.

Due to his motivation to identify and repatriate deceased border-crossers, Martín Zamora became a well-known man in and beyond Spain, especially in the Tadla-Azilal region in the Middle Atlas of Morocco, from where many people have emigrated since the 1990s. He has gained an ambiguous fame over the last 15 years. Local and national newspapers have published articles about him and his business, and a feature movie has been made based on his story. His profession and his dedication to repatriate deceased border-crossers have not only evoked support, but also sceptical voices. In my interviews, representatives of NGOs criticised the specialisation and professionalism of his business as a way to earn money and representatives of Moroccan migrant associations disapproved of the elevated costs for post-mortem repatriations. As mentioned before, many families could not cover the expenses and thus Martín Zamora tried, sometimes very successfully, to receive payment from different state institutions.

Business with the dead is inherently ambiguous and a person who makes his or her living from the dead tends to become suspicious. Martín Zamora himself is very aware of this ambiguity. Sometimes people allege that he is pleased about the death of a person – an insinuation that he vehemently rejects: 'To be good in what you are doing doesn't mean to be happy when someone dies!' Indeed, he became very personally affected by the extreme conditions of death in a dispossessed context and during the interviews he repeats various times: 'Everybody has his own story, every dead has his own story'. Martín Zamora not only listened to these stories, he lived them with the families and he became part of them. The tragedies of unauthorised border crossings and the mourning of the survivors touched him: 'I got involved in every way, I even became a Muslim'. And he acknowledges that only a Muslim is allowed to touch a dead Muslim, as his work entails. Yet even if he did not say it explicitly, it seems that the appreciation he received once he identified and repatriated a dead person was overwhelming and spiritually fulfilling. In addition to that, he has become an important and above all trusted contact for those crossing the border:

> Sometimes they called me from a *patera*[5] for help, from the *patera*, because they feared that it would sink. I had to inform the Guardia Civil and they asked me, how I knew. I told them, that they called me, they were probably family of a dead one and maybe somebody told them: 'Look, write down this phone number, this man will help you' … and immigrants came to the *tanatorio* asking for help … we helped somehow, in other words, we got involved, very, very involved.

He did not seek or ask for this involvement, but he was torn between personal attachment and business, and thus it just somehow happened to him. The responsibility has been tremendous and when I ask him about the limits of his engagement, he simply answers: 'There are no limits', as if he did not have another choice: 'You live their stories with them, they become somehow a part of yourself'.

Listening to Martín Zamora, it is remarkable that he primarily uses the words 'person' (*persona*) and 'the dead' (*muerto/s*) when he talks about the deceased and he only speaks about 'bodies' and 'corpses' when he explicitly refers to his business or to forensic examinations. As I mentioned before, our speech reveals something about our attachments and detachments and how we ethically relate to the dead. Thus, it is not surprising when talking about his business that he unconsciously reinforces his identity as a businessman and creates an emotional distance by relating in a rather calculating manner to dead bodies. Through Martín Zamora's story we can observe how an undertaker gets involved in the political, economic, and emotional dimension of migratory death and how he has become a connecting link between the dead body and the bereaved.

Doing fieldwork in a border zone where death under extreme conditions occurs and seeking a conversation with the living, with those who have successfully crossed the border as well as with the bereaved, the ethnographer might also become a connecting link. The anthropologist and director of a human rights organisation Reinecke (2013) outlines her search for a missing husband and father by illustrating excluding power mechanisms of identification systems and technologies, bureaucratic obstacles, and errors in data collection for families of deceased migrants. Thanks to her, the bereaved finally gained certainty over the death and could bury the dead. During fieldwork, I myself have become a source of knowledge for families regarding burials, resting places, memorials, and the confirmation of the certainty of death. The undertaker and the ethnographer investigate the death of a person, both seek to know something about the deceased's life, both listen with empathy to the stories of the survivors and bereaved, they enter a relationship with them and seek to learn about their experiences and perceptions. The undertaker and the ethnographer walk a line between personal attachment and self-interest; and often they enter into professional relationships, which soon become friendships.

Conclusions

In this article, I have focused on various actors who are involved in the post-mortem reception of deceased border-crossers. I have shown divergent forms of attachment and detachment among these actors (including the researcher) who are – willingly or not – connected to border-related deaths. My experiences in talking with 'experts of death' and their sensitive and respectful way of speaking have led me to the assumption that their work does not turn into an indifferent routine. A dead person – known or unknown – matters. Nevertheless, the ethical question of how one should proceed with dead border-crossers remains unanswered.

To conclude, I wish to highlight three points I have made in the article and which I consider to be essential regarding the uncertain and diverse belongings of the dead. First, exploring the junction of death and belonging (being and longing), I have advocated a semantic awareness of the depersonalising quality of words such as cadaver and corpse that deprive the dead of their belonging. At the same time I have shown the importance of the presence of the dead body for mourning processes and have argued that a dead body – due to its absence – becomes a subject of longing and desire. Rites of passage and especially rites of reintegration aim to reunite the living and the dead but if the bereaved do not know about the death of a loved one, rituals cannot be performed and thus the transition from life to death is not accompanied in the proper cultural and religious ways.

Second, I have shown that the non-identification and anonymity of dead border-crossers and thus the uncertainty of their belongings are not fixed but produced in different ways by different local actors and the state itself. The uncertainty might be generated, concealed, or rewritten. Further, I have argued that the moral claim of equality in the treatments of dead bodies disguises their multiple belongings and the lack of an Islamic cemetery is not only a disappointing reality for many Muslims living in the Campo de Gibraltar, but it is also a source of resentments. Intercultural and interreligious negotiations in the border area could enable diversified and thus more dignified burial practices and forms of commemoration.

Third, I have highlighted the ambiguity of the term belongings, which includes the meaning of belonging as religious, emotional, genealogical, and national attachments and identities, as well as the personal possessions which travel with migrants. These personal things tell us something about the dead person and they are sometimes successfully used to trace the identities of the dead. Thus, personal belongings have the power to reconstitute the identity of a person.

To deepen our understanding of different ways of identification at the border, we need to ask what identification actually means. To do so, we should seek to know who a person was during his or her lifetime; to figure out where he or she was coming from or heading to, to reveal where and to whom a person belongs. In transnational settings, the question of where a person belongs cannot be easily answered. Family and friends might be dispersed across national borders and the person may have more than one home. In this context the question *to whom* a person belongs becomes much more relevant. Not trusting the testimony of a relative, but relying exclusively on biological proof is a violent act that leaves family and friends left behind with feelings of powerlessness and impotence. It is my view that the political, ethical, and academic demand for identification and post-mortem repatriation or transfer has to develop a broader understanding of the complexity of multiple belongings and identification processes itself by creating an awareness of power relations in the context of the treatment, management, and administration of dead bodies at the border.

Acknowledgements

I would like to thank Eva Soom Ammann for inviting me to contribute to this special issue and Alistair Hunter for his thorough reading and constructive comments on the paper. He and Sabine Strasser shared essential thoughts with me on the ambiguity of the term belongings, I am truly grateful for that. I would like to extend thanks to the two anonymous reviewers and the Intimate Uncertainties project team: Luisa Piart as well as Veronika Siegl and Julia Rehsmann for feedback on earlier versions of this paper. I also wish to express my gratitude to Driss el Hadj and to the participants of my research.

Notes

1. A database of numbers derived from official sources is available at http://www.borderdeaths.org/ (see, Last 2015) and the Andalusian human rights organization APDHA publishes annual reports with death estimates: http://www.apdha.org/.
2. This study is embedded in a larger research project called "Intimate Uncertainties. Precarious Life and Moral Economy across European Borders" (https://intimateuncertainties.wordpress.com/), which seeks to understand the making of moral worlds and the governing of precarious lives in transnational circuits marked by social, political, legal, and economic inequalities. The project is directed by Sabine Strasser, University of Bern. This work was supported by the Swiss National Science Foundation (149368).
3. Although Hasse has the German etymology in mind, I believe that his analysis can be easily applied to English and Spanish.
4. Pseudonym.
5. *Patera* is the Spanish word for a small, wooden boat. Since the emergence of irregular migration towards Spain via the Mediterranean Sea and Atlantic Ocean, any boat used by border crossers is now called *patera*. In the 1980s until the beginnings of 2000, border crossers actually used wooden boats. Nowadays they generally use *zodiacs*, big inflatable dinghies with a motor, to cross the Alboran Sea or *toys*, small inflatable dinghies with paddles, to cross the Strait of Gibraltar.

References

Alscher, S., 2005. *Knocking at the doors of "fortress Europe": migration and border control in Southern Spain and Eastern Poland*. Working paper 126. *The centre for comparative immigration studies CCIS*. San Diego: University of California.

Andersson, R., 2014. *Illegality, Inc. Clandestine migration and the business of bordering Europe*. Oakland: University of California Press.

Basch, L., Glick Schiller, N., and Blanc-Szanton, C., 1994. *Nations unbound. Transnational projects, postcolonial predicaments, and deterritorialized nation-states*. New York: Gordon and Breach.

Beck, J., *et al.*, 2015. Animal scavenging and scattering and the implications for documenting the deaths of undocumented border crossers in the Sonoran dessert. *Journal of forensic science*, 60, 11–20.

Butler, J., 2004. *Precarious life: the powers of mourning and violence*. London: Verso.

Ferrer-Gallardo, X., 2008. The Spanish-Moroccan border complex: processes of geopolitical, functional and symbolic rebordering. *Political geography*, 27, 301–321.

Grant, S., 2011. Recording and identifying European frontier deaths. *European journal of migration and law*, 13, 135–156.

Grant, S., 2015. Migrant deaths at sea. Addressing the information deficit. *Migration policy practice*, 5 (1), 9–16.

Hasse, J., (2016). *Versunkene Seelen. Begräbnisplätze ertrunkener Seeleute im 19. Jahrhundert*. Freiburg and München: Herder

Jackson, M., 2013. *The wherewithal of life. Ethics, migration, and the question of well-being*. Berkeley: University of California Press.

Last, T., 2015. Deaths at the borders: database for the southern EU, 12 May. Available from: http://www.borderdeaths.org/wp-content/uploads/Press-Release-EN.pdf [Accessed 10 October 2015].

Last, T. and Spijkerboer, T., 2014. Tracking deaths in the Mediterranean. *In*: T. Brian and F. Laczko, eds. *Fatal journeys. Tracking lives lost during migration*. Geneva: IOM (International Organization for Migration), 85–106.

Macho, T., 1987. *Todesmetaphern. ZurLogik der Grenzerfahrung*. Frankfurt am Main: Suhrkamp.

Macho, T., 2000. Tod und Trauer im kulturwissenschaftlichen Vergleich. *In*: J. Assmann, ed. *Der Tod als Thema der Kulturtheorie. Todesbilder und Totenriten imalten Ägypten*. Frankfurt am Main: Suhrkamp, 89–120.

Magaña, R., 2008. *Bodies on the line: life, death, and authority on the Arizona-Mexico*. Chicago: Proquest UMI Dissertation Services.

Magaña, R., 2011. Dead bodies. The deadly display of Mexican border politics. *In*: F.E. Mascia-Lees, ed. *A companion to the anthropology of the body and embodiment*. West Sussex: Wiley-Blackwell, 157–171

Marcus, G.E., 1995. Ethnography in/of the world system: the emergence of multi-sited ethnography. *Annual review of anthropology*, 24, 95–117.

Moorehead, C., 2014. Missing in Mediterranean. *Intelligent Life Magazine. May–June*. Available from: http://moreintelligentlife.com/content/features/caroline-moorehead/lost-mediterranean?page = full [Accessed 10 October 2015].

Probyn, E., 1996. *Outside belongings*. New York: Routledge.

Rauf, H., 2014. *The final journey: what to do when a Muslim passes away in Scotland*. Glasgow: Glasgow Central Mosque.

Regan, M., 2010. *Death of Josseline: immigration stories from the Arizona borderlands*. Boston: Beacon Press.

Reineke, R., 2013. Lost in the system: unidentified bodies on the border .*NACLA report on the Americas*, 46 (2), 50–53.

Samaoli, O., 2007. *Retraite et vieillesse des immigrés en France*. Paris: L'Harmattan.

Sharp, L., 2009. The commodification of the body and its parts. *Annual review of anthropology*, 29, 287–328.

Strasser, S., 2009a. *Bewegte Zugehörigkeiten. Nationale Spannungen, transnationale Praktiken und transversale Politik*. Vienna: Turia & Kant.

Strasser, S., 2009b. Transnationale Studien: Beiträgejenseits von Assimilation und "Super-Diversität". *In*: M. Six-Hohenbalken and J. Tošić, eds. *Anthropologie der Migration. Theoretische Grundlagen und interdisziplinäre Aspekte*. Vienna: Facultas Verlags- und Buchhandels AG, 70–92.

Strasser, S., 2009c. Europe's other. Nationalism, transnationals and contested images of Turkey in Austria. *European societies*, 10 (2), 177–195.

Tarrés, S., *et al*. 2012. *Migrar, morir, ¿retornar? Un programa de investigaciónsobre la muerte en context migratorio*. Bilbao: *VII Congreso de Migraciones internacionales en España*, 11–13 April.

Verdery, K., 1999. *The political lives of dead bodies: reburial and postsocialist change*. New York: Columbia University Press.

Weiss-Krejci, E., 2013. The unburied dead. *In*: S. Tarlow and L. Nilsson Stutz, eds. *The oxford handbook of the archaeology of death and burial*. Oxford: University Press, 281–301.

Zagaria, V., 2011. *Grave situations – the biopolitics and memory of the tombs of unknown migrants in the Agrigento province*. (MSc). London School of Economics and Political Science.

Index

absence of the absence 103–5
absent mourning 101–115; *see also* uncertain belongings
acceptance 25–6
acculturation 69
aged care in institutions 27–8
aging immigrants 69–73
Ahorn-Grieneisen 55, 62
amplified analysis 42
ancestors category 92
anticipating cross-cultural interaction 13–19
art of endurance 24–38
assimilation 73, 95–7
Association of Friends and Families of the Victims of Clandestine Immigration 104
associations between religious indicators 79
assumptions 19–21
Assyrian Christianity 88–98
asymmetrical dynamics of power 64
attachment 2, 54, 69, 87, 93, 96, 111–13
Attias-Donfut, C. 87, 94
attitudes and practices 75–8; individual characteristics 77–8
autonomy 2, 26–7, 31, 33, 36
awareness of power relations 112–13

'bad death' 2–3; *see also* 'good death'
Baile, W.F. 10–11, 21
balancing money and safety 90–91
Balint, M. 10–11
Becker, H.S. 49
becoming an undertaker 55–6
being at peace 93–4
belonging 2, 100–115
bereaved family perspective on end-of-life care 42–5
bereavement support 53–4
Berger, J. 1
between business and engagement 108–112
between civil society and state 53–67
beyond end-of-life care 39–52; bereaved family perspective 42–5; care providers' perspective 45–7; conclusion 48–50; methods 41–2; research population 42; structural features 47–8; transitions in care situations 40–41
bio-psychological entireties 11
biomedicine 25
border control 101–115
bureaucratic competence 53–67
burial at sea 56
burial location preferences 85–100; conclusion 97–8; integration through burial 95–7; methods 88–90; motivations for 90–95
burial repatriations 101–115
burial Spanish style 105–8
business 108–112
Butler, J. 105

cancer 26, 31–2
care logics *see* logics of care
care providers' perspective on end-of-life care 45–7
Chaïb, Y. 87, 95
challenge of 'doing good death' 30–35; guilt-ridden dying trajectory 31–3; indecisive dying trajectory 34–5; uneased dying trajectory 30–31; unprotected dying trajectory 33–4
challenged 'death work' 35–7
challenges of cross-cultural interaction 13–19
characteristics influencing attitudes to funeral practice 77–8
Christian migrants 85–100
citizenship 13
claiming land, faith, family 85–100
classical assimilation theory 73
Clayton, J. 86
co-construction of 'good death' 24–38
codified religious ritual practices 94
collaboration 45
collective identity 71
common experiences 47–8
communication challenges 15–20, 26, 39–40, 47–50, 63, 103
compartmentalised society 49–50
competence of bureaucracy 58–61
concept of 'good death' 25–6
conceptualisations of stereotyping 20–21

concrete 'death work' 36
conflict resolution 53–4
congruence 36
constructivism 12
contradictory goals 24–38
convenience sampling 12
coordination 90–91
coping with contradictory goals 24–38
coping strategies 70–72
Coptic Christianity 88–98
countering stereotypes 61–4
credibility 56, 61
cremation 54–5, 87
Cresswell, T. 86
cross-cultural interaction 9–23; expecting difference 14–15; facing dilemmas 15–17; misunderstanding 17–19
culpability 107
cultural differences 48–50
cultural experts 33
cultural mediation 53–67
culturalisation of patients 10
culture competent care 4, 9–11, 20–21
curbing negative perceptions 61–4

data on religion 73–4
death and end-of-life rituals 5–6
'death work' 24–5, 28–30, 35–7; challenged 35–7
death-care business 55–6, 58
deathscapes 86
deciding about life and death 35
deciphering Otherness 20–21; *see also* Otherness
dehydration 29, 34–5
dementia 29, 33–5
descendants category 92–3
descendants of Turkish migrants 58–84
detachment 61
deterioration 32
Deutsche Islam Konferenz 60
deviance 20–21
diaspora 59, 98
dichotomisation of traditions 71–2, 79
difference 14–15, 20, 60, 64, 96
different religious indicators 79
differential allocation of grievability 105
dignity 25–7, 36
dilemmas 15–17
DIMECCE project 88–9, 94
discrimination 62–3, 72
discursive violence 103
disenfranchisement 110
diversity and migration 13, 26
DNA samples 104, 110
'doing diversity' 27–8
domination through knowledge 64–5; *see also* cultural mediation
drowning 107, 110
dual nationality 93
dying and end-of-life care 4–5
dying trajectories 25–35; *see also* challenge of 'doing good death'
'dying well' 27

economic concerns 90
emotional distance 104
emotional landscape 93–4
end-of-life care 4–5
end-of-life rituals 5–6
endurance 24–38
engagement 108–112
Epner, D.E. 10–11, 21
erasure of difference 64
eschatological doctrine 95
Esser, H. 96
ethnic sameness 37
ethno-cultural diversity 10–11, 13, 15, 20–21
euthanasia 25
expecting difference 14–15
experts of death 112–13
explicit knowledge 26
extension of responsibility 15–17, 20
external border of EU 101–115

facing dilemmas 15–17
facing God 31, 40
family chain migration 70, 72
family considerations in burial 91–3; ancestors category 92; descendants category 92–3; nearest and dearest category 92; parents category 92
fear of dying 18
Federal Association of German Undertakers 55
feeling at home 93–4
feeling uncertainty 14–15
fertility levels 70
financial costs 90
first ancestors 91–3
flattening differences 60
focus group interviewing 11–13
full disclosure 19
funerary practice 69–73, 77–8; characteristics influencing attitudes 77–8

Gardner, K. 87
genealogical continuity 91–3
genealogical reconstruction 108
Generations and Gender Survey 69, 73–4
geo-sociological elements 95–6
German burial restrictions 56
Germany: Muslims in Berlin 53–67; Turkish migrants and descendants 68–84
GGS *see* Generations and Gender Survey
Giddens, A. 26
'going native' 95–7
'good death' 2–3, 24–38; art of endurance 35–7; challenge of 'doing' 30–35; 'death work' 28–30; in institutional aged care 27–8; negotiations of diversity 26; notions of 25–6
'good enough death' 28
Grant, S. 104, 109

grave re-use 94–6, 107
grief psychology 55
Grounded Theory 27–8
Guardia Civil 109
Guba, E.G. 12
guesthood 87
guilt-ridden dying trajectory 31–3
Gunaratnam, Y. 10, 36
Gysels, M. 11

Hasse, J. 103–4
Heaton, J. 42
heterogeneity 3, 27, 70
holism 10–11
homeland 89–91, 96–7
Howarth, G. 61
human rights 104
human trafficking 58
hyperthermia 107

illness-centred medicine 10–11
impact of socio-demographic characteristics 79–80
implicit knowledge 26
importance of religious funeral ceremony 68–84; aging immigrants and religiosity 69–73; associations between religious indicators 79; conclusion 79–80; data on religiosity 73–4; individual characteristics 77–8; method 74–5; religious attitudes among migrants 75–7
importance of socio-demographic characteristics 68–84
incurable illness 40–42
indecisive dying trajectory 34–5
Indigenous Australians 86
individual identity 71
individualisation 25
institutional aged care 27–8
institutional 'good death' 24–38
insurance 41, 48
integration through burial 69, 77–8, 95–7
inter-religious dialogue 60
intercultural negotiations 53–67; *see also* Islamic undertakers
intergenerational tensions 42–4, 49, 70
interment of remains 88
Interpol 109
Iraqi Christianity 88–98
Islamic deathways project 57–8
Islamic undertakers 53–67; becoming an undertaker 55–6; bureaucratic competence 58–61; conclusion 64–5; countering stereotypes 61–4; method 57–8
isolation 42–3
issues of 'doing death' 27–8
'Italianness' 33

Jackson, M. 103
Jones, K. 10
Jonker, G. 71–2, 80

Kai, J. 10
Knijn, T. 49

lack of common language 18, 20; *see also* communication challenges
land/faith/family 85–100
language difficulties 15–19, 47–9, 63
Last Judgment 88
Last, T. 102
legal–institutional barriers 2
legitimacy of lived religion 95
life expectancy 70
life-support 25, 35
Lincoln, Y.S. 12
lines of collision 4, 25–6, 35–7
living up to expectations 14–15
living wills 29
location of Other 20–21
logics of care 49
logistic regression models 74–5
longevity 27

Macho, T. 103–4
made-to-measure care 48
Magaña, R. 103–4, 106–7
maintaining life 2; *see also* prolongation of life
making assumptions 19–21
manageability of dying 2
Marjavaara, R. 87
maximum-variation sampling 12
meaningful identification 86, 97–8
Mecca 40–41, 56, 71, 73, 107
memorialisation 54, 86–7
Middle East 85–100
migrant backgrounds 57–8, 75–7
miscommunication 18; *see also* communication challenges
misunderstanding 17–19
mobility of the dead 54, 86–7
modernisation 25
Modood, T. 96
Mohr, J. 1
morphine 31–2
mortuary sciences 55
moving from ill to deceased 42–7; bereaved family perspective 42–5; care providers' family perspective 45–7
multi-morbidity 27
multiple belongings 101–115
Muslim migrants 53–84; Turkish 68–84; as undertakers 53–67

narratives of nation 93
National Board of Health and Welfare (Sweden) 13
'natural death' 27, 29
naturalistic inquiry 12
nearest and dearest category 92
negotiations of diversity 26
Netherlands 39–52

networking 103
'new first ancestors' 91–3
nichos 101–2, 105
9/11 62–3
non-marital cohabitation 75
non-self-determined dying 36
notions of 'good death' 25–6

Oliver, C. 95
ontological insecurity 36, 90
ontological security 3, 12, 21, 26, 36–7
ontologising tendency 103
orchestration of funerary practices 45, 50
organisational costs 91
orthopraxy 98
Otherness 4, 6, 9–23; challenge of cross-cultural interaction 13–19; conclusion 19–21; methods 11–13; patient-centredness 10–11; Sweden 13
outsourcing dying 25

palliative care 3–4, 10–12, 21, 25–30, 35, 39–49
parents category 92
passports of the dead 58, 63
patient-centredness 10–11
patienthood 17, 20
perception of norm 14, 17
Perloff, R.M. 10
Phalet, K. 73
Pickering, M. 20
pioneering spirit 42–6
pious indifference 94–5
place-making 86
plurality 36
political afterlife 104
political mediation 64–5; *see also* cultural mediation
post-migration difference 6, 96
post-mortem repatriation 41, 54, 71–2, 87, 91, 101–115
power relations 48–50, 64, 112–13
practical considerations in burial 90–91; financial costs 90; organisational costs 91; security situation 91
pre-death transitions 2
prejudice 5, 15, 61–2, 64
preponderant presence 93, 96–7
presence of an absence 103; *see also* absence of an absence
present bodies 103–5
problems about unmet needs 17–19
professionalism 33–4, 36–7, 49–50, 61, 111
prognosis 26
prolongation of life 25–6, 32
Proshansky, H.M. 86
Protestantism 30–31, 37
provider understandings of patients 9–23; *see also* Otherness
psychological burdens 41
psychosocial needs 26
purposeful inscription of space 94–5

quality of life 35, 39–40
quantitative studies 73–4
Qur'an 40, 45

racial profiling 62
Ramadan 46, 73
Rauf, H. 107
reception of deceased migrants 112–13
Regan, M. 104
Reimers, E. 86–7
Reinecke, R. 104, 112
reintegration 112–13
relatives' roles 43
religiosity 69–74; data on 73–4; and funerary practice 69–73
religious affiliations 73–4, 94–5
religious attitudes 75–7
religious conformity 94, 96
religious counselling 53–4
religious disassociation 94
reorientation of identity 95–7
repatriation 41, 54, 71–2, 91, 102–4; *see also* post-mortem repatriation
research populations 42
responsibility 15–17
rigorous religious argumentation 37
rites of passage 112–13
ritual washing 41, 48, 71, 105
ritualisation of death 54
role of cultural mediation 64–5; *see also* cultural mediation
role of socio-demographic characteristics 68–84
routines of 'doing death' 24–5
rupture of communication 103

sacralising aspirations 95–6, 98
sacred considerations in burial 94–5; religious conformity 94; religious disassociation 94–5; sacralising aspirations 95
safety and coordination 90–91
Samaoli, O. 107–8
sameness 4; *see also* Otherness
Schengen Area 102, 107
Scholl, I. 11
Schulz, F. 56
secondary analysis of data 41–2
secularisation 73, 77–8
securitised border zones 3
security situation 91, 96, 98
sedation 40
sedative pain therapy 31
selective secularisation 73
Self vs Other 20–21
self-conception 87
self-determined dying 2, 25–7, 29, 31
self-presentation 62–3
self-reported health 13
self-understanding 92
Selman, L. 11
semi-structured interviews 12, 88–90

sepulchral culture 56
Shi'a Islam 57
SIVE 102
Smits, F. 75
social cohesion 70
social exclusion 72
socio-demographic characteristics 68–84
socioeconomic integration 79
solidarity 86
sovereignty 55
Spain 101–115
Spanish undertakers 108–112
spatial fix 86
special needs 10
speechlessness of dead 106
Spijkerboer, T. 102
spiritual needs 26
staking a claim to land 85–100
stereotyping 15, 20–21, 61–4
stigmatisation 53, 61
Strasser, S. 103
striking a balance 90–91
structural features of end-of-life care 47–50
Strumpen, S. 70
studies into dying and death 1–8; death and end-of-life rituals 5–6; dying and end-of-life care 4–5
sub-Saharan countries 101–115
subjective religiosity 72–3
Sudnow, D. 28
suffering 29, 32, 35
suffocation 107
suicide 25
Sunni Islam 46, 57, 71, 107–8
surveillance 102
Sweden 9–23
Swiss Ethnological Society 28
Swiss National Science Foundation 24–5
Switzerland 24–38

taphonomy 104
terminal illness *see* incurable illness
terminal phases 29–33, 42
territorial considerations in burial 93–4; emotional landscape 93–4; narratives of nation 93; preponderant presence 93
theory of place-identity 86
Thompson, W. 61
ties of belonging 2
'timing' death 25
Tonkinson, M. 86
total institution 28, 36
tranquillity 93–4
transitions in care situations 39–41, 45–7
transnational ageing 85
transparency 61
Tricks of the Trade 49
Turkish Funeral Funds 71, 79–80
Turkish migrants 68–84
typology of motivations 90–95; family considerations 91–3; practical considerations 90–91; sacred considerations 94–5; territorial considerations 93–4

UN Human Rights Council 104
uncertain belongings 101–115; between business and engagement 108–112; burial the Spanish way 105–8; conclusion 112–13; present bodies 103–5
uncertainty 14–15
understandings of migrant backgrounds 9–23; *see also* Otherness
undertakers 55–67, 108–112; *see also* Islamic undertakers; Spanish undertakers
uneased dying trajectory 30–31
uniqueness 4, 11, 21
unknown, the 13–15
unmet needs 17–19
unprotected dying trajectory 33–4
unsuccessful negotiation 26, 36

variables on religion 73–4
Verdery, K. 106
Verhagen, S. 49
village mentality 59–60
vulnerability 85

Weber, M. 64
WHO *see* World Health Organisation
whole body inhumation 88
Wolff, F.C. 87, 94
World Health Organisation 11–12, 26
World War II 1
worry about facing dilemmas 15–17

Zamora, Martín 108–112

For Product Safety Concerns and Information please contact our EU
representative GPSR@taylorandfrancis.com
Taylor & Francis Verlag GmbH, Kaufingerstraße 24, 80331 München, Germany

www.ingramcontent.com/pod-product-compliance
Ingram Content Group UK Ltd.
Pitfield, Milton Keynes, MK11 3LW, UK
UKHW051830180425
457613UK00022B/1186